OCT   2015

ALSO BY JON PALFREMAN

The Case of the Frozen Addicts (coauthor)
The Dream Machine (coauthor)

BRAIN STORMS

BRAIN STORMS

THE RACE TO UNLOCK
THE MYSTERIES OF
PARKINSON'S DISEASE

JON PALFREMAN

SCIENTIFIC AMERICAN / FARRAR, STRAUS AND GIROUX

NEW YORK

Scientific American / Farrar, Straus and Giroux
18 West 18th Street, New York 10011

An excerpt from *Brain Storms* originally appeared, in slightly
different form, in *Scientific American Mind*.

Library of Congress Cataloging-in-Publication Data
Palfreman, Jon, author.
    [Brain storms (Palfreman)]
    Brain storms : the race to unlock the mysteries of Parkinson's disease /
Jon Palfreman. — First edition.
        p. ; cm.
    Includes index.
    ISBN 978-0-374-11617-0 (hardback) — ISBN 978-0-374-71185-6 (e-book)
    I. Title.
    [DNLM: 1. Parkinson Disease—Personal Narratives. WL 359]
RC382
616.8'33—dc23
                                                                2015003861

Designed by Jonathan D. Lippincott

Scientific American / Farrar, Straus and Giroux books may be purchased for
educational, business, or promotional use. For information on bulk purchases,
please contact the Macmillan Corporate and Premium Sales Department at
1-800-221-7945, extension 5442, or write to specialmarkets@macmillan.com.

www.fsgbooks.com • books.scientificamerican.com
www.twitter.com/fsgbooks • www.facebook.com/fsgbooks

Scientific American is a registered trademark of Nature America, Inc.

1   3   5   7   9   10   8   6   4   2

CONTENTS

AUTHOR'S NOTE

I could not have written this book without the generous advice and assistance of numerous researchers, clinicians, professional colleagues, and fellow people with Parkinson's. My thanks to the many individuals who shared their knowledge, stories, and advice in interviews and e-mails, including Chuck Adler, Jay Alberts, Ron Alterman, Krys Bankiewicz, Roger Barker, Carol Ann Bassett, Sara Batya, Tom Beach, Anders Björklund, Bastiaan Bloem, Paul Bolam, DuBois Bowman, Debi Brooks, Patrik Brundin, James Buie, Jean Burns, Jacqueline Burré, Jerry Callahan, Doug Carnine, Linda Carnine, Cuiping "Tracy" Chen, Chad Christine, Daniel Corcos, Mahlon DeLong, Dennis Dickson, Chris Dobson, Roger Duvoisin, Nancy Egan, Stanley Fahn, Matt Farrer, Jeremiah Favara, Richard Fisher, Michael Fitzgerald, Kim Gannon, Steve Gill, Christopher Goetz, Larry Golbe, Tom Graboys, Dan Grossman, Katrina Gwinn, Zach Hall, John Hardy, Joel Havemann, Dustin Heldman, Hampus Hillerstrom, Fay Horak, Tom Isaacs, Ole Isacson, Dave Iverson, Matt Johnson, Karl Kieburtz, Tuomas Knowles, Jeff Kordower, Walter Koroshetz, Mark Kramer, Rajaraman Krishnan, Nell Lake, Story Landis, Bill Langston, Peter Laufer, Virginia Lee, Andrew Lees, Dan Leventhal, Peter LeWitt, Oded Lieberman, Max Little, Walter Maetzler, Chester Mathis, Helen Matthews, Pietro Mazzoni, James

McNames, Karsten Melcher, August Moretti, Robert Nussbaum, Jay Nutt, Michael Okun, Ben Petrick, Elizabeth "Eli" Pollard, Mihael Polymeropoulos, Pamela Quinn, Sara Riggare, Web Ross, John Rothwell, Joan Samuelson, Pete Schmidt, Todd Sherer, Ludy Shih, Jerry Silbert, Jonathan Solomon, Maria Spillantini, Jon Stamford, Phil Starr, Georg Sternberg, Farah Stockman, A. Jon Stoessl, Carlie Tanner, Barbara Tilley, Alex Tizon, Michele Vendruscolo, Monica Volz, Karen Weintraub, Christine Woodside, Zbigniew Wszolek, and Richard Wyse. Unless otherwise cited in the notes, all quotations are derived from these interviews.

In a few places, names and identifying characteristics have been altered to protect the privacy of individuals.

BRAIN STORMS

PROLOGUE

In June 2012, I visited the eminent neuroscientist Bill Langston at his home in Los Altos Hills, California. We had met a quarter of a century before, when I had produced a documentary for the PBS series *Nova* titled "The Case of the Frozen Addict." The film told the story of six young drug abusers in San Jose, California, mysteriously struck with the symptoms of Parkinson's disease, a neurodegenerative condition that normally affects the elderly. Bill Langston, then an unknown clinician, discovered the unlucky individuals languishing in psych wards and jail cells and had temporarily reversed their symptoms with the drug L-dopa, the principal treatment prescribed for Parkinson's disease.

Over the next few months, Langston and his colleagues cracked the mystery. The young people, it turned out, had injected a bad batch of synthetic "designer" heroin. Unfortunately for them, the backstreet chemist who'd concocted the drug had made a terrible mistake during the synthesis and inadvertently created a neurotoxic contaminant called MPTP.

While tragic for the victims, this deadly molecule proved of immense scientific importance. Scientists had been hampered in their efforts to study Parkinson's because humans are the only animals to contract it naturally. To make real progress, scientists need ways to study diseases in animals—using an "animal model."

MPTP, therefore, changed everything for scientists interested in Parkinson's disease. This neurotoxin, it turned out, could rapidly induce parkinsonism in monkeys, as it had in the six addicts. As Bill Langston put it, MPTP was "a bracing tonic . . . Suddenly we had ways to study why cells die in Parkinson's disease. With the animal model, we could test new medicines as fast as you could make them."

Langston went on to become an internationally renowned neuroscientist, founding his own research institute in Sunnyvale, California: the Parkinson's Institute and Clinical Center. My film established my reputation as a documentary producer and science journalist.

That afternoon in Los Altos Hills, Bill and I spent three hours talking about some of the remarkable advances in Parkinson's research that had taken place over the past two decades. The conversation ranged from neural grafting to gene therapy, from novel drugs to therapeutic vaccines, from genetics to neurosurgery. It was fascinating. But this time, ironically, I wasn't there just as a journalist. I had a compelling personal reason for my renewed interest in Parkinson's disease. At the age of sixty, I'd contracted the condition myself.

I'd learned my destiny on a cold, dark, rainy January morning in 2011. My doctor, worried about a modest tremor in my left hand, had referred me to the Oregon Health and Science University's movement disorder center in Portland. I wasn't particularly worried. My mother had manifested a similar symptom most of her life, caused by a relatively benign condition called essential tremor. I was convinced that the same diagnosis would explain my shaking. A young neurologist named Seth Kareus examined me. After taking my medical history, Seth conducted a series of clinical tests. He asked me to perform various motor tasks—rotating my hands back and forth, touching my index finger and thumb together, tapping my foot up and down as quickly as I could. He palpated my muscles in all four limbs, checking for both strength and tone. He tested my reflexes with a reflex hammer. He watched as I tracked a moving object with my eyes. He asked me to walk up and down the hall.

After twenty minutes of tests, he told me I had Parkinson's disease. My symptoms were, so far, mild and confined to the left side of my body, but the malady would inevitably progress, and in time I would need treatment with drugs.

I left the hospital in a state of shock. It took me more than a year to process this news, a year in which I engaged in a series of coping strategies. There was secrecy: the only person I told during the first three months after my diagnosis was my wife. There was denial. I questioned the diagnosis and consulted other neurologists. They confirmed I had Parkinson's disease. There was self-pity. And there was isolation. I didn't reach out to other Parkinson's sufferers. To the contrary, I wanted nothing to do with them. The fragile, bent, trembling figures I observed in neurologists' waiting rooms saddened and angered me. Was this really who I would become?

Gradually, I began to think more clearly. Because there was no denying my condition, it made sense to embrace it. I threw myself into reading everything I could about Parkinson's disease and speaking to neuroscientists and clinicians so that I could understand my predicament. That's what had brought me down to Los Altos Hills in June 2012 to mine Bill Langston's encyclopedic grasp of the field. After all, as a lifelong science journalist who had reported on this disease, I was better placed than most to figure out the state of Parkinson's research and ascertain what kind of future I faced. In a profound sense, understanding Parkinson's disease and finding a cure are now my journalistic beat.

·

Parkinson's isn't new. Its odd symptoms have been noticed throughout history. An Egyptian papyrus from the twelfth century B.C. describes an old king who drooled (a symptom of advanced Parkinson's disease). The ancient Indian Ayurvedic texts mention a progressive condition characterized by tremor and lack of movement. And the Greek physician Galen even distinguished two kinds of parkinsonian tremors: resting tremors and action tremors. Leonardo da Vinci observed people "whose soul cannot control their movements in spite of the fact that their extremities are shaking continuously." There is a clear reference to the disease in Shakespeare's *Henry VI, Part 2* when Dick the butcher asks, "Why dost thou quiver, man?" To which Lord Say replies, "The palsy, and not fear, provokes me." And scholars agree that the seventeenth-century philosopher Thomas Hobbes almost certainly suffered from it.

But it was James Parkinson who gave the first full clinical description of the malady that bears his name in his 1817 *Essay on the Shaking Palsy*. Parkinson based his remarkable discourse on six cases he had either examined as patients or observed when walking around his London neighborhood. His words—stunning in their acuity—echo across two centuries. "The first symptoms perceived," writes Parkinson, "are, a slight sense of weakness, with a proneness to trembling in some particular part; sometimes in the head, but most commonly in one of the hands and arms." This is the resting tremor that I, and many other sufferers, manage on a day-to-day basis.

Here's some of what we know about this unusual disease in 2015: About seven million people globally and one million Americans have Parkinson's, with some sixty thousand new U.S. cases each year. It's more common in men than women (except in Japan), and the prevalence increases with age: among people over eighty years old, one in fifty will get the disease. Remarkably, it's less common among smokers and coffee drinkers. Everybody knows someone with Parkinson's disease, and many famous people—including the former attorney general Janet Reno, the Intel executive Andy Grove, the evangelist Billy Graham, the singer Johnny Cash, the British soccer icon Ray Kennedy, the Olympic runner John Walker, the actor Michael J. Fox, the comedian Robin Williams, the writer Martin Cruz Smith, the rock star Linda Ronstadt, and the boxer Muhammad Ali—have developed the malady. Conventional wisdom holds that the disease is relentlessly progressive. Often starting, as James Parkinson noted, with a tremor affecting one limb on one side of the body, the condition typically spreads to all four limbs. The patient's body becomes more rigid, frequently leading to a stooped posture, and movements slow down and become smaller and less fluid. As the disease advances—usually slowly—the patient becomes more and more disabled.

As I began to accept my fate and read more about my disease, my questions became more focused and carried a sense of urgency. Can Parkinson's be slowed, stopped, or even reversed? Can the disease be prevented before it starts, like polio and smallpox? Perhaps. Two hundred years after Parkinson's essay, people still live and die with his disease. But today, more than at any time in history, success seems possible. For this is the age of neuroscience. Having sequenced the human genome, biomedical researchers have their sights set on the ultimate frontier—the human brain. And in that quest—including such projects as the $300 million, ten-year BRAIN (Brain Research Through Advancing Innovative Neurotechnologies) initiative, which seeks to map the activity of every neuron in the human brain—many scientists view Parkinson's as a pathfinder disease, perhaps the best hope of making sense of an organ of mind-boggling complexity. In Parkinson's, the automatic acts we take for granted, like swinging our arms when we walk, start to break down, and this breakdown opens a window into the brain. The scientific challenge is to bridge an awesome difference of scale—to connect the abnormal movements of Parkinson's patients to exotic processes going on in the brain involving tiny molecules that weigh a billionth of a trillionth of an ounce.

Brain Storms tells the story of a disease that has entranced physicians and scientists for two centuries, from the first reports of its symptoms to the front lines of biomedical research, where we may be on the cusp of a major advance. It represents my journey—both as a journalist and as a patient—examining the latest discoveries about this fascinating and insidious neuro-degenerative curse and evaluating the prospects for a cure. It's a story with many setbacks, not the least being the revelation that the classic motor symptoms of Parkinson's disease—tremor, rigidity, slowness of movement, and postural imbalance—may be the tip of a clinical iceberg. It now seems that Parkinson's disease takes hold of an individual decades before any tremors appear and continues wreaking damage throughout the brain until the end of life. This means that in addition to movement problems, people with Parkinson's, or Parkies, as we sometimes call ourselves, have to cope with a wide set of adverse symptoms from constipation to dementia.

But from that setback has come a new theory of the disease. The bad actor, many researchers argue, may be a common protein called alpha-synuclein, which goes rogue, forming sticky toxic aggregates that jump from cell to cell inside the brain, killing neurons as they go. Since similar rogue proteins are found in Alzheimer's disease (the most common form of dementia) and in other currently incurable neurodegenerative conditions such as Huntington's disease and Creutzfeldt-Jakob disease, some scientists foresee the possibility that one day in the not-too-distant future they'll be able to protect against all neurodegenerative diseases, preventing these scourges from ever happening.

As I've learned, people with Parkinson's progressively lose core pieces of themselves. We forget how to walk. Our arm muscles get weaker. Our movements slow down. Our hands fumble simple fine-motor tasks like buttoning a shirt or balancing spaghetti on a fork. Our faces no longer express emotions. Our voices lose volume and clarity. Our minds, in time, may lose their sharpness . . . and more.

No one wants Parkinson's. But there are many worse fates. Unlike many cancer victims, people with Parkinson's tend to survive for a long time. During this period, we (unlike our more cognitively impaired counterparts with Alzheimer's and Huntington's disease) can report lucidly on our condition until the end. Our insights can help unpack the disease and assist in the scientific pursuit of better therapies and ultimate cures. As I have learned from fellow sufferers, there are many ways to fight back against this disease. From the Boston cardiologist Thomas Graboys to the NBA player Brian Grant, from the dancer Pamela Quinn to the engineer Sara Riggare, these fellow Parkinson's sufferers have inspired me with their courage, resilience, ingenuity, and wisdom. And I'll be highlighting many of their stories in this book.

The brain is an awesomely resilient organ. As neurologists have found, islands of ability remain even in the most impaired individuals. Take the weird phenomenon of kinesia paradoxica, for example. Advanced Parkinson's cases will frequently freeze up when walking, their feet effectively glued to the floor. But remarkably, the simple act of drawing a line on the ground (or placing a foot) in front of them can interrupt this "gait freezing"; the Parkinson's patient will step over the line (or foot) and carry on her way. Equally paradoxical, and equally fascinating, is the finding that Parkies too disabled to walk can still ice-skate, ride a bike, or run. Then there are some truly astonishing studies in which certain patients in the placebo arms of clinical trials have markedly reversed their parkinsonian symptoms for long periods of time. Such findings indicate that the brain has many tricks that may offer novel therapeutic possibilities—ones that neuroscientists have so far missed.

The conquest of neurodegenerative diseases is everyone's business, for one simple reason: the world is getting older. By 2050, the number of people aged sixty-five or over will triple. These 1.5 billion elderly people—16 percent of the world's population—will be at risk for brain diseases such as Parkinson's and Alzheimer's. There's no time to waste.

1

"DISCOVERY"

The disease, respecting which the present inquiry is made, is of a nature highly afflictive . . . whilst the unhappy sufferer has considered it as an evil, from the domination of which he had no prospect of escape.

—James Parkinson, *An Essay on the Shaking Palsy*, 1817

Every three years, people with Parkinson's gather for a Parkinson's world summit of sorts. Over four days, patients—desperate for a cure—rub shoulders with biomedical professionals who have devoted their lives to conquering this disease. In 2013, at the third such world congress, in Montreal, the major Parkinson's charities, such as the Michael J. Fox Foundation, the American Parkinson Disease Association, the National Parkinson Foundation, Parkinson's UK, the Cure Parkinson's Trust, and the Parkinson's Disease Foundation, are present in force. So are pretty much all of the world's leading Parkinson's researchers, eager to report on their latest efforts to understand and defeat this awful malady. And there are lots of patients—some like me, who so far have only mild symptoms, and others who move around with canes, walkers, or wheelchairs.

The opening ceremony is intensely moving. United by a common enemy, more than three thousand Parkinson's patients, caregivers, researchers, and clinicians from sixty-five countries come together in Montreal's giant Palais des Congrès. The audience is animated, frequently rising to give standing ovations for the inspiring speakers whose images are displayed on massive video screens. The mood is defiant. Speaker after speaker urges the congregation not to give up hope.

I find it energizing to see so many Parkies in one place. We are in fact a rather impressive group, which includes accomplished dancers, musicians, filmmakers, entrepreneurs, scientists, and even true-life heroes, like the former NASA astronaut Rich Clifford and the endurance athlete Alex Flynn. The fact that so many people here have achieved great success in their professional lives supports my belief that if we all work together with the scientists, we can beat this disease.

The star performer that first night is the Canadian celebrity Tim Hague Sr. A Parkinson's patient since 2011, Hague (together with his son, Tim Hague Jr.) has against all odds just won the reality show *The Amazing Race Canada*, where pairs of contestants race across the length and breadth of Canada. On September 16, 2013, Canadian viewers watched as Tim and his son beat out the competition in the first season of the show. The story of his extraordinary physical triumph serves as a potent reminder of the power of the human spirit.

Hague steps up to the microphone and addresses all the stakeholders. "Whether you're the researcher, the health-care professional, the family member, the friend, the person with Parkinson's, persevere . . . our journey together may very well be a long one . . . don't give up. You never know what is just around the next corner . . . never lose hope . . . persevere, this race as well can be won."

Over the next four days, the attendees show no sign of giving up. Researchers try not to dwell on bad news and sometimes play down the considerable difficulties of (and delays in) bringing scientific advances to the clinic. Scientists and patients alike freely use the "cure" word. Even though I know much of it is exuberance, for those four days I willingly buy into this conspiracy of hope. I too feel the sense of urgency. After all, Parkinson's is now my world—a world where it is crucial to believe tomorrow can be better than today.

As the dozens of scientific sessions show, the quest to defeat Parkinson's is a long, complicated war being fought on many fronts. Some researchers focus on understanding the disease in detail. This basic research—often carried out with test tubes and laboratory rodents—generates lots of ideas about potential weak spots, so-called targets where a drug or other intervention might just slow, stop, or reverse the progress of the disease. But only a few of these ideas get out of the lab and into clinical trials, where other researchers test the new agents to see if they're safe and effective. There are geneticists searching for clues in patients' genomes. There are epidemiologists looking for factors that appear to increase or decrease the likelihood of contracting the disease, from pesticides, which appear to increase the risk of Parkinson's, to coffee and smoking, which seem to be protective against getting it. There are clinicians focused on understanding the varied symptoms of patients, and there are neuropathologists who study the cellular damage to the body and brain after death. And more.

A social scientist observing all the delegates sporting their colorful badges would surmise that we all shared a brand. In our case, it's not a product like Coca-Cola or Nike but a commitment to defeat a disease named after a man who lived some two hundred years ago. The observer would wonder as I had about the man behind the brand. Who was James Parkinson, and how did he come to have a disease named after him? And what is it about this condition that so fascinates the world of science and medicine? This is where our story of Parkinson's really begins.

·

There are no portraits or engravings of James Parkinson,* just a
commemorative stone tablet in St. Leonard's church, Shoreditch,
the East London parish where he was born, lived, and died. His
home has long since been demolished. If you go to Shoreditch,
you'll see that a simple plaque adorns the building that now stands
at 1 Hoxton Square, the address where he ran his medical practice.
The son of a physician, Parkinson studied Latin, Greek, natural
philosophy, medicine, surgery, drawing, and shorthand; the latter
he considered essential for taking medical histories. His published
writings reveal the eclectic interests of an Enlightenment intel-
lectual. In addition to producing medical discourses on mental
illness, gout, hernia, and appendicitis, he dabbled in chemistry, pa-
leontology, and political activism—writing pamphlets under the
nom de plume Old Hubert.

Parkinson seems to have been a curious and compassionate
man. He would almost certainly be unknown today, however,
but for one thing: a small monograph he published in 1817 about
a new disease he called the shaking palsy. A careful observer, he'd
noticed the syndrome during his regular walks around the streets
of East London. From time to time, he came across people who
moved differently from the crowd, and when he saw them, he'd
approach them for an interview to find out more. One day, for
example, he encountered a sixty-two-year-old man who had
worked as an attendant in a magistrate's court. His body, observed
Parkinson, was "much bowed and shaken. He walked almost
entirely on the fore part of his feet, and would have fallen every
step if he had not been supported by his stick." The attendant
told Parkinson that he was resigned to "the incurable nature of

*A photograph published in a book called *Medical Classics* (1937–38) is sometimes mis-
taken for James Parkinson. In fact, it's of a dentist with the same name. The "real" James
Parkinson died in 1824, fifteen years before the earliest form of photography—produced by
the daguerreotype process—was developed.

his complaint." Based on just six cases, only two of which he examined fully, Parkinson came up with a description of what he suggested might be a new disease.

An Essay on the Shaking Palsy is a beautiful piece of medical literature, one that people with Parkinson's everywhere will recognize captures much of what they go through—including tremor, poverty of movement (also called bradykinesia), and postural instability. "Walking," Parkinson wrote, "becomes a task which cannot be performed without considerable attention. The legs are not raised to that height, or with that promptitude which the will directs, so that the utmost care is necessary to prevent frequent falls."

Despite its brilliance, few physicians noticed Parkinson's essay. As a consequence, nineteenth-century individuals struck down with the condition were left to figure out matters on their own. One of the most moving stories that I have come across is that of the Prussian linguist, diplomat, and educational reformer Wilhelm von Humboldt (1767–1835). Ignorant of Parkinson's essay, Humboldt documented his own parkinsonian decline in a series of wrenching letters to a friend, Charlotte Diede. He noted his stooped posture, complained of "an intolerable slowness and clumsiness" when unbuttoning clothes, and reported that his handwriting was shrinking (what's now called micrographia). In a poignant passage written on November 4, 1833, he laments, "Every line starts, with best intentions, in large letters only to end, with ill success, in barely legible small ones. If my life hadn't taught me patience and self-control, this would seem to me insupportable." Not understanding that he had a neurodegenerative disease, Humboldt interpreted his symptoms as accelerated aging following the death of his wife. After seven years of Parkinson's symptoms, Humboldt died of pneumonia.

Humboldt joins a list of intellectuals in history who suffered with the symptoms of Parkinson's disease before the infirmity had been recognized and named. Another was the seventeenth-century

English philosopher Thomas Hobbes. John Aubrey wrote in his *Brief Lives* that Hobbes "had the shaking palsy in his hands; which began in France before the year 1650, and has grown upon him by degrees, ever since, so that he has not been able to write very legibly since 1665 or 1666, as I find by some of his letters to me."

The disease that James Parkinson noticed would gain widespread recognition thanks to the nineteenth-century French physician Jean-Martin Charcot, the second major figure in the history of this disease. In his day, Charcot, a short, stocky figure with a striking resemblance to Napoleon, was a medical celebrity. According to the neuroscientist and historian Christopher Goetz, people came from all over the world to watch Charcot's clinical lectures at the Salpêtrière Hospital in Paris. Housing five thousand patients, three thousand of whom had neurological conditions, the Salpêtrière was a neurologist's paradise. Whereas James Parkinson had informally looked at just six cases with one common syndrome, Charcot methodically analyzed hundreds of patients with a wide range of odd disorders. He soon discovered several neurologically distinct entities—including multiple sclerosis, amyotrophic lateral sclerosis, Charcot-Marie-Tooth disease (a peripheral nervous system disorder involving loss of touch sensation), and the shaking palsy.

James Parkinson's 1817 essay was hardly known in France. But sometime in the 1860s, Charcot obtained a copy and immediately realized its importance. By carefully observing his own patients at the Salpêtrière, he codified (more precisely than Parkinson) the disease's four common symptoms—tremor, rigidity, slowness or poverty of movement, and postural instability—and added two more, which Parkinson had missed: small handwriting (the micrographia that von Humboldt had noticed) and facial masking (hypomimia), in which the patient's facial expression is lost or diminished because of altered muscle tone.

The perceptive Charcot—whose students included Sigmund Freud and William James—noticed that not all patients had tremor (about one in five patients lacked this symptom). Charcot argued that given this fact, calling the condition the shaking palsy was misleading. He proposed instead the label "Parkinson's disease." And it stuck.

By the 1880s, thanks to his extensive clinical research at the Salpêtrière, Charcot had essentially completed the clinical picture of Parkinson's disease, at least when it came to the motor symptoms. He would have had little difficulty distinguishing those of us at the Palais des Congrès in Montreal who had Parkinson's from those who didn't. And in addition to becoming an expert at diagnosing the condition, he started treating patients' symptoms, like tremor, with plant-based formulations that he came up with by trial and error. He prescribed hyoscyamine—an extract of jimsonweed—in pill form rolled into bits of white bread. Other medicines were derived from belladonna (deadly nightshade).

Charcot developed other intriguing therapies. Having observed that the symptoms of Parkinson's patients appeared to improve after long rides in carriages, in trains, and even on horseback, he speculated that the vibrations might be therapeutic. So Charcot developed an electrically powered "shaking chair," or *fauteuil trépidant*. One of his students, Gilles de la Tourette, refined this concept into a portable shaking helmet that vibrated the brain. His therapeutic vibration concept was recently tested in a controlled trial using commercially available massage chairs. Patients were assigned to one of two groups: one cohort had daily sessions in a vibrating chair for one month; the other had the same number of sessions in the same chair but with the vibration switched off. Both groups were exposed to relaxing natural sounds, such as ocean waves. The researchers concluded that what Charcot observed was largely a placebo effect, in which perceived benefit had more to do with the patient's and the clinician's wishful expectations of improvement than the vibrational

Figure 1: Jean-Martin Charcot's shaking chair (*fauteuil trépidant*), designed to relieve the symptoms of Parkinson's patients (Reprinted with permission from *Cold Spring Harbor Perspectives in Medicine* 2011;1:a008862, copyright © Cold Spring Harbor Press, 2011)

Figure 2: This vibrating helmet, designed by Charcot's student Gilles de la Tourette, applied the shaking directly to the patient's brain. (Reprinted with permission from *Cold Spring Harbor Perspectives in Medicine* 2011;1:a008862, copyright © Cold Spring Harbor Press, 2011)

therapy. (And it turns out that the placebo effect, which I'll discuss in chapters 8 and 16, plays an important role in Parkinson's disease.)

We can thank Jean-Martin Charcot, then, for clinically defining, naming, and even attempting to treat the disease that had brought me to Montreal that week in 2013. In reality, however, in Charcot's day, Parkinson's was not yet a disease in the true sense of the word but merely a cluster of symptoms or, in medical parlance, a "syndrome." Before a syndrome can be classified as a true disease, physicians need to possess at least one of two additional pieces of knowledge: how the malady started or how it ends. Knowing a syndrome's *cause* is the clearest sign you have a real disease—as occurred when scientists discovered that the human immunodeficiency virus (HIV) caused acquired immunodeficiency syndrome (AIDS). If a cause is unknown, then physicians and scientists hunt for specific pathological changes in the patient's tissues, which, in the case of the brain, are usually detected in a postmortem examination after the patient dies.

In the 1880s, scientists had little idea what caused parkinsonism, but pathologists routinely dissected patients' brains looking for signs of damage to various tissues. Charcot, originally trained as a pathologist, taught the "anatomo-clinical" method, which sought to connect the clinical features of a disease like Parkinson's to anatomical changes or "lesions" in the brain.

In a typical postmortem dissection at the Salpêtrière, a pathologist peeled back the face, cut open the top of the skull, and removed the brain (quite similar to how a postmortem dissection would be performed today). But thereafter, all he had to guide him was the gross appearance of the brain's "white" and "gray" matter. The upper portion of the brain—the cerebral hemispheres—looks a bit like the cap of a mushroom, and the rest of the brain resembles its stem. The mushroom cap is split by a deep canyon (dividing the left and right hemispheres) and covered by a wrinkled outer layer of gray matter called the "cortex" (the Latin word for

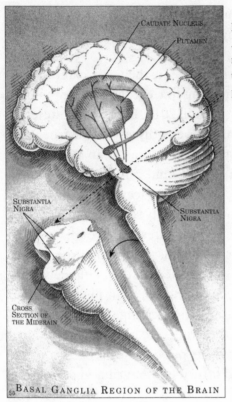

CAUDATE NUCLEUS

PUTAMEN

SUBSTANTIA NIGRA

SUBSTANTIA NIGRA

CROSS SECTION OF THE MIDBRAIN

BASAL GANGLIA REGION OF THE BRAIN

Figure 3: A view of the basal ganglia. Axons from the substantia nigra, a major source of dopamine, extend to the striatum, which is made up of the putamen and the caudate nucleus.

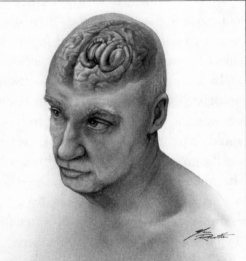

Figure 4: Another view of the basal ganglia region of the brain, showing the putamen and the caudate nucleus

(Copyright © Marie Rossettie, CMI)

"bark"). Underneath the cortex is largely white matter punctuated with islands of gray matter. When Charcot's pathologists sawed into the brain horizontally (slicing from the top to the bottom) or coronally (going from the back to the front), they observed different brain structures, which earlier anatomists had assigned Latin names—ventricles (bellies), corpus callosum (tough body), corpus striatum (striped body), globus pallidus (pale globe), thalamus (inner chamber), and substantia nigra (black stuff). This last structure was so named because its cells contained the pigment melanin, making them dark.

The anatomo-clinical method depended on finding differences between the brains of healthy and sick individuals. Because most impoverished Salpêtrière patients were wards of the state, autopsies were commonplace, so Charcot and his colleagues had numerous opportunities to find odd lesions in deceased patients' brains that might explain the neurological symptoms they suffered in life. One day in 1893 (the same year that Charcot died), two of Charcot's students, Paul Blocq and Georges Marinesco, got a lucky break. They admitted to the Salpêtrière a thirty-eight-year-old patient with a parkinsonian tremor and rigidity on the left side of his body. The patient subsequently died of pulmonary complications and was autopsied. The postmortem turned up a hazelnut-sized lump in the right side of his midbrain, very close to the substantia nigra. Blocq and Marinesco's discovery inspired Édouard Brissaud, Charcot's successor at the Salpêtrière, to hypothesize that the substantia nigra—the site of the black stuff—was the key pathological source of Parkinson's disease.

Was this a meaningful finding or just coincidence? For twenty-five years, no one bothered to systematically investigate the matter. Then, in 1919, Constantin Tretiakoff, a Russian graduate student working in Paris, offered convincing proof that the black stuff was indeed associated with Parkinson's disease. His dissertation examined fifty-four autopsied brains, nine of which had Parkinson's. Tretiakoff found that all the genuinely parkinsonian brains showed

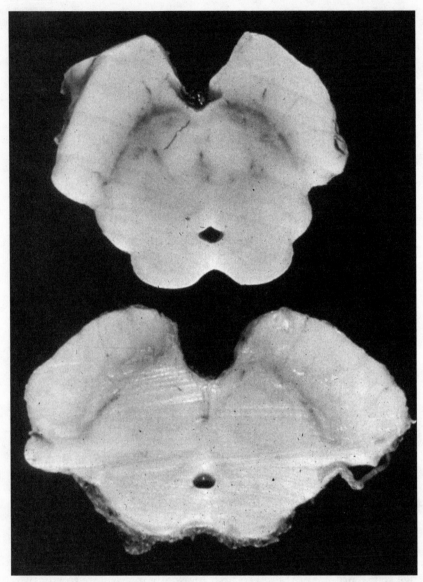

Figure 5: Section of a healthy substantia nigra with pigment, compared with one from a Parkinson's patient (Ronald C. Kim, MD)

extensive damage to the substantia nigra, whereas none of the healthy controls did. As Figure 5 shows, the difference between normal and parkinsonian brains could scarcely have been clearer: the diseased brains had simply lost their black stuff.

Tretiakoff also noticed something else. Some of the brain cells of deceased Parkinson's patients contained small spherical structures. They were roughly the size of a red blood cell and were surrounded by a clear halo. Tretiakoff named them "corps de Lewy," or "Lewy bodies," to acknowledge their discoverer, Frederick Lewy, a German pathologist working in Dr. Alois Alzheimer's Munich laboratory.

Pathologists and neurologists would in time come to accept that Lewy bodies were *the* defining hallmark of genuine Parkinson's disease. Neurologists might diagnose parkinsonism in life. But

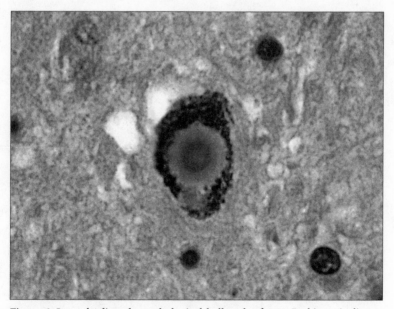

Figure 6: Lewy bodies, the pathological hallmark of true Parkinson's disease
(Image courtesy of Mark L. Cohen, MD, and FrontalCortex.com)

only after pathologists found Lewy bodies in a patient's damaged substantia nigra after death would they be sure that he truly had Parkinson's disease.

While scientists still had no idea of what caused Parkinson's, it could now fairly be called a disease rather than a syndrome. By the 1950s, more than 130 years after James Parkinson's essay, scientists—especially Charcot, Tretiakoff, and Lewy—had established the commonly known features of the Parkinson's brand. This brand framed the disease as a "movement disorder," mostly found in the elderly, and resulting from damage to a very small region of the brain—the substantia nigra.

So far, little had been found to effectively treat the condition, but that was about to change, thanks to the efforts of a group of brilliant Swedish and Austrian researchers. They proved that it wasn't only the black stuff that vanished when neurons died in Parkinson's patients. Something else disappeared as well—a brain chemical called dopamine.

2

RESTORATION

Bedridden patients who were unable to sit up, patients who could not stand up from a sitting position . . . performed all these activities with ease . . . they walked around . . . they could even run and jump . . . This dopa effect reached its peak within two to three hours and lasted, in diminishing intensity, for 24 hours.

—Oleh Hornykiewicz and Walther Birkmayer, 1961

Roger Duvoisin, a major figure in Parkinson's research and the author of a popular book, *Parkinson's Disease: A Guide for Patient and Family*, is one of the few surviving members of an elite group of neuroscientists that fifty years ago pioneered a therapeutic revolution. He recently sent me a riveting piece of video that documents what life was like for Parkinson's patients before and after this transformation. He remembers the period in the late 1960s when the video was made as one of the most exciting times in his professional life.

Originally shot on 16-millimeter film, the video shows Duvoisin as a young physician examining one of his early Parkinson's patients—labeled patient 3—on January 17, 1968. The opening scene, depicting the situation before the medical revolution, is bleak. We see a head-and-shoulders view of a roughly sixty-year-old man wearing polka-dot pajamas, sitting on a chair at a clinic in the Columbia-Presbyterian Medical Center in New York. His face lacks all expression, and his arms shake with a violent tremor. The camera cuts to a side view, and we see from the way he is sitting that the man is effectively frozen in place, quite unable to stand up. A youthful-looking Duvoisin then pulls patient 3 to a standing position, but the patient appears to have no sense of balance. Duvoisin needs to support him like a giant marionette. Suddenly Duvoisin lets go, and the patient starts falling back into the chair—until Duvoisin's arm rescues him. It's a devastating clinical picture. This man, only six years after diagnosis, is dying of Parkinson's.

But in 1968, for the first time since James Parkinson published his *Essay on the Shaking Palsy*, neurologists and their patients had reason to hope. For in this year, as the film later goes on to document, clinicians like Duvoisin were experimenting with a new medicine called levodopa, or L-dopa.

The story of this revolution can be traced back to the late 1950s, to some ingenious experiments on rabbits. In 1957, the Swedish scientist Arvid Carlsson became interested in the recently discovered antipsychotic drug reserpine. When he administered sufficient doses of this drug to animals, he found it temporarily paralyzed them. Carlsson speculated that reserpine blocked the uptake in the brain of a key neurotransmitter and that the resulting chemical imbalance led to drug-induced Parkinson's-like symptoms. On a hunch, he tried to pharmacologically reverse the rabbits' somnolence with L-dopa. Carlsson expected the L-dopa to pass into the animals' brains and be converted into a class of chemicals called catecholamines (notably the neurotransmitters adrenaline and noradrenaline), which would restore the chemical balance.

His hunch was correct. The L-dopa worked. Almost magically, the rabbits awakened, became alert, and started to move normally. But when he examined the balance of chemicals in their brains, Carlsson discovered that the L-dopa had been converted not into noradrenaline, as expected, but rather into a different chemical called dopamine. At the time, most neuroscientists regarded dopamine as an unimportant molecule, a mere chemical building block for manufacturing other molecules rather than an important neurotransmitter in its own right. But Carlsson now realized that this view was misguided. He became convinced that dopamine was an important neurotransmitter without which the brain could not function. In a lecture at the National Institutes of Health (NIH) the following year, Carlsson theorized that a dopamine deficiency might be the neurochemical basis of Parkinson's disease in human beings.

While many scientists were skeptical, two Austrian researchers, Oleh Hornykiewicz and Herbert Ehringer, were inspired by Carlsson's suggestion. They examined autopsied human brains and found that, as Carlsson had argued, the neurons in the striatum (a brain region thought to be critical for the orchestration of movement) of deceased Parkinson's patients had virtually no dopamine. Hornykiewicz then teamed up with the physician and researcher Walther Birkmayer and proved that dopamine was missing from the substantia nigra region as well—the site of the black stuff that was typically deficient in Parkinson's patients' damaged brains.

The work of Carlsson, Hornykiewicz, Ehringer, and Birkmayer formed the basis of a new dopamine-centered theory of Parkinson's, the foundation for a therapeutic revolution that Duvoisin and other neuroscientists (including the neurologist Melvin D. Yahr) would build. The theory of how Parkinson's disease developed went as follows: First, for reasons that were not clear, a sizable fraction of a person's dopamine-producing neurons in the substantia nigra dies off; these neurons, neuroanatomists knew, normally send long fibers (some two centimeters long) called axons to dispatch dopamine to other nerve cells in the striatum.

Second, because of the death of dopamine-producing neurons in the substantia nigra, less and less dopamine travels along those axons and reaches its target in the striatum. Third, deprived of sufficient dopamine, the elaborate orchestration between brain and muscles starts to break down. And finally, the characteristic symptoms experienced by cases like patient 3 appear—tremor, slowness, rigidity, loss of facial expression, gait and balance problems.

Mercifully, the new dopamine theory suggested a possible treatment for such patients. Because the parkinsonian brain is low on dopamine, the logical step is to replace the missing chemical. But getting chemicals into the brain is not as simple as taking a pill, the researchers realized. The brain is protected by its "blood-brain barrier"*—a tightly woven network of cells designed to prevent potentially harmful chemicals from entering the cerebral sanctum. The same barrier, it turns out, blocks certain neurotransmitters like dopamine from directly reaching the brain. Instead, the brain comes equipped to manufacture its own dopamine from chemical building blocks like L-dopa, which *can* pass through the blood-brain barrier—as Carlsson had shown with his rabbit experiments. So in 1961, unencumbered by today's institutional review boards and regulatory agencies, Birkmayer and Hornykiewicz tried administering small quantities of L-dopa intravenously into twenty patients with advanced Parkinson's disease.

Just as L-dopa had slipped through the blood-brain barrier, been converted to dopamine, and "awakened" Carlsson's rabbits, the Austrian researchers found that it worked similarly on human patients, at least for a few hours. Hornykiewicz recalls, "It was a spectacular moment to see the patients who could not walk, could

---

*In the 1880s, the German bacteriologist Paul Ehrlich observed that injecting water-soluble dyes (for example, trypan blue) into animals stained all their organs except the brain and the spinal cord. In 1913, Edwin Goldman, Ehrlich's student, demonstrated that the very same dyes, when injected directly into the cerebrospinal fluid, readily stained nervous tissue but not other tissues.

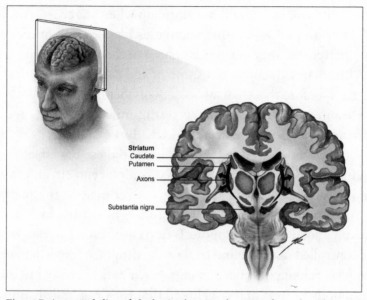

Figure 7: A coronal slice of the brain showing the axons from the substantia nigra extending to the striatum (Copyright © Marie Rossettie, CMI)

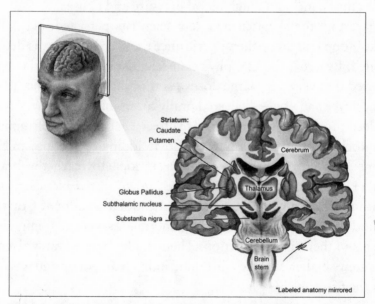

Figure 8: A coronal slice of the brain showing the parts of the basal ganglia (Copyright © Marie Rossettie, CMI)

not get up from bed, could not stand up when seated, start walking. They all performed these activities like normal. Speech became better . . . they started laughing and actually crying with joy." Birkmayer documented some of the cases in a film that he presented before the Medical Society of Vienna.

Despite such extraordinary testimonials, Duvoisin recalls that many scientists were unconvinced: "Birkmayer was using tiny amounts of L-dopa—a few milligrams. So these big effects were puzzling . . . people wondered if it was a placebo effect." It was also known that much of the L-dopa never made it to the brain but was broken down by enzymes in the bloodstream. Careful research added to doubts about L-dopa's efficacy. In 1966, in the first controlled double-blind trial of the drug, the Swedish neurologist Clas Fehling examined twenty-seven patients who'd received intravenous L-dopa and concluded that the drug had absolutely no effect on Parkinson's symptoms when compared with saline solution. Worse, one-third of patients suffered serious dopa-related side effects, including high blood pressure and nausea.

Despite the skepticism, a few scientists persisted, believing that L-dopa might be therapeutic once the correct dose was determined. By 1967, the U.S. physician-scientist George Cotzias had proved that very, very large doses—a thousand times higher than Birkmayer had used—delivered orally for months at a time, could indeed produce anti-parkinsonian effects. Moreover, Hoffmann–La Roche chemists had discovered that adding carbidopa, an enzyme blocker, enabled more of the L-dopa to reach the brain rather than being chemically broken down in the bloodstream. With this new dosing regimen for carbidopa-levodopa, Cotzias found that a group of eighteen patients made spectacular improvements in their motor function. Their rigidity melted away. Tremors grew smaller. Wheelchair-bound individuals got up and walked for the first time in years.

Duvoisin treated his first case with carbidopa-levodopa in December 1967. Like Cotzias, he was quickly convinced: "The effect was so dramatic I couldn't believe it. Patients were so improved that they didn't look like they had Parkinson's anymore."

To document what he saw, Duvoisin shot the 16-millimeter films of his first series of patients treated with L-dopa. Which brings us back to the video of patient 3. On January 17, 1968, patient 3 had been rigid, like a statue. In the film that Duvoisin shot on March 20, 1968, five weeks after starting levodopa therapy, patient 3 looks strikingly changed. The camera starts with the same head-and-shoulders shot, but this time patient 3 is smiling expressively, with a sparkle in his eye. The movie cuts to the side view of him sitting relaxed on the chair. This time, he stands up effortlessly by himself and begins to walk with perfect arm swings toward the clinician, at which point he makes a 180-degree turn and walks back across the room. When standing still, he has no problem balancing, and when Duvoisin pushes him to test his balance response, patient 3 takes a small step back and recovers without difficulty. The final scene shows the patient at a blackboard, where he takes a stick of chalk and fluently writes out the words "This is a lovely day." A simple chemical restores a frozen man to an almost normal state.

Anyone watching this metamorphosis would think of L-dopa as powerful . . . even magical. It's easy to see why neurologists embraced it so eagerly (Duvoisin reports that clinicians treated several thousand patients experimentally in the next few years) and why regulatory bodies like the Food and Drug Administration (FDA) quickly approved Merck's Sinemet and Roche's Madopar (the brand names for their L-dopa formulations) for the routine treatment of Parkinson's.*

Few episodes in medical history are as dramatic as the discovery of L-dopa. Neuroscientists speak about it in almost reverential

---

*To enable more of the L-dopa to reach the brain, Merck's Sinemet adds carbidopa, which inhibits the enzyme dopa decarboxylase in the peripheral blood. The name Sinemet comes from the Latin phrase *sine emit*, meaning "no vomiting." A Hoffmann–La Roche drug, Madopar, combines levodopa with a different dopa decarboxylase inhibitor, benserazide.

terms. The University of Rochester Medical Center's Karl Kieburtz, an authority on the history of the drug, describes L-dopa as "one of the most potent therapies in all of neurology—indeed in all of medicine—think about it: to take someone who was essentially rigid like a stone . . . and enable them to get out and walk and function . . . it's unbelievable . . . and we almost missed it."

But clinicians soon discovered that after a "honeymoon period" L-dopa exacted a price for the miracle. Cotzias and Duvoisin noticed that within a month or two of starting the drug, their newly mobile patients displayed new disabling motor side effects (quite different from the nausea and high blood pressure mentioned earlier), in particular, involuntary writhing movements of the arms, legs, or head called dyskinesias (Greek for "bad movements"). Doctors also observed that over time the drug became less and less therapeutic and had to be taken in higher doses and more frequently. Most bothersome, patients reported so-called on-off fluctuations, in which the medicine's power to combat Parkinson's disease suddenly vanished without warning. As Duvoisin recalls, "The patient would be walking around and suddenly she would be frozen. One patient said the experience was like flicking a light switch on the wall. And so the phenomenon became known as the 'on-off effect.'" Finally, some patients suffered side effects like confusion, agitation, paranoia, and hallucinations. Neurologists euphemistically called all these L-dopa-induced side effects "motor complications."

The ecstasy and the agony of L-dopa were dramatically captured by a young neurologist at Beth Abraham Hospital in the Bronx named Oliver Sacks. In 1969, Sacks decided to give L-dopa to a group of eighty people who had survived a mysterious outbreak of encephalitis lethargica in the 1920s. This was a Parkinson's-like condition that had rendered its victims so unable to move that it was sometimes referred to as the sleeping sickness. As Sacks movingly describes in the first edition of his book *Awakenings* in 1973, he administered massive doses of L-dopa, and at first he saw great improvement in these patients who had been frozen for forty years. As he writes, "The patient on L-DOPA enjoys a perfection of being, an ease of movement and feeling and thought, a harmony of relation within and without." But it was only temporary. Within a short time, Sacks noted, the patient's "happy state—his world—starts to crack, slip, break down, and crumble; he lapses from his happy state, and moves toward perversion and decay."

While L-dopa was vastly superior to what came before, the drug fell far short of being a cure. On the one hand, the L-dopa allowed "frozen" wheelchair-bound individuals to walk again and increased patients' life expectancy. On the other hand, virtually all patients taking levodopa were sentenced to future disabling motor complications. And that's as true today as it was in 1970.

It's perhaps understandable, given L-dopa's mix of positive and negative attributes, that some newly diagnosed patients, like me, are initially reluctant to start using it. Viewed from the patient's perspective, L-dopa therapy seems like a Faustian bargain: You trade a better "now" for a worse "later." You exchange moving well in the immediate future for complications—like moving too much and suddenly switching off—down the road.

Because of this uncertainty about L-dopa, many neurologists start patients on a less powerful class of drugs called dopamine agonists, medicines that mimic the action of dopamine; two common ones are Requip and Mirapex. Discovered in the 1970s, these drugs work by "pretending" to be dopamine. They pass freely through the blood-brain barrier and actually trick the receptors in the striatum into action. While the brain isn't actually receiving dopamine, it "thinks" it is and reacts accordingly. Patients and clinicians will tell you that they are only about half as effective as L-dopa and they have their own set of unpleasant side effects, ranging from nausea to sleep attacks to compulsions. But I decided to start out with dopamine agonists and transition after a year to L-dopa. And like millions of patients before me, the L-dopa restored some of my lost self. My stiffness vanished, my tremor almost disappeared, and for much of the time an observer would be hard-pressed to notice anything unusual about the way I moved.

People with Parkinson's are very lucky to have L-dopa. There is no equivalent therapy for other neurodegenerative conditions like Alzheimer's, Huntington's, or Lou Gehrig's disease. Whatever its limitations, L-dopa turned Parkinson's from a condition in which victims experienced a rapid slide toward immobility and death into a chronic disease with a gradual trajectory of decline. A consequence is that within a decade of Cotzias's pioneering work, there were hundreds of thousands of people with Parkinson's living functional lives and surviving ten, fifteen, twenty years, and more. These individuals began to associate and advocate for their disease.

By the time I was diagnosed in 2011, most areas of the United States had thriving patient support groups. I've found them to be fascinating ways not only to get to know what most concerns my fellow patients but also to receive practical advice. Since my diagnosis, I've visited several groups, including one on the East Coast that meets once a month on a Sunday afternoon. About twenty people, two-thirds of them men, participate in the sessions. Most are highly educated individuals in their sixties. While the mood of the meetings is usually positive, some people express sadness and anxiety about their altered fates.

At the start of one meeting in 2013, the facilitator asked us to introduce ourselves. After we gave our names, all participants followed group custom by identifying themselves using the same metric: the time since we were diagnosed. I realized while I was announcing myself to the group that I was conceding something profound: that the diagnosis marked an irreversible change in my identity, the moment that one version of me ended and another version began.

I looked around the room and noticed quite a variation in clinical presentation. There were people with mild symptoms who'd had the disease for ten years and folks who looked seriously impaired who'd had it for less than two. Some people shook with serious tremors, and some didn't. Some said they were well enough to work full-time; some reported they'd retired and were drawing Social Security Disability Insurance.

The conversation ranged from sophisticated exchanges about prescription drugs and their side effects—about which some group members knew quite a lot—to credulous discussions about alternative therapies like fish oil and vitamins. Monica, a woman in her fifties, mentioned a 2013 FDA report of a possible link between Mirapex (the dopamine agonist I had started on) and heart failure. And this generated a wider discussion of the dopamine agonist drug class. Two group members, a retired dentist and an accountant, warned the group that compulsive personality types should be wary of Mirapex. As the dentist, Phil, told us, "The Mirapex made me feel better, but it also made me feel disinhibited." He'd developed a gambling compulsion, which put his family savings at risk. The accountant, Tom, reported a sex addiction that caused him to start having affairs.

Initially, when I had heard about this side effect of dopamine agonists, I didn't believe it. After speaking with five different clinicians about their experience and reading a series of clinical studies on the links between compulsive behavior and dopamine agonists like Mirapex, I've changed my mind. The research shows that as many as one in ten patients are susceptible to impulse-control disorder (ICD) brought on by dopamine agonists and to a lesser extent by L-dopa. Dopamine, after all, plays a central role in the brain's reward and pleasure system, so it's not really all that surprising.

I spoke to the dentist, Phil, after the meeting in which he raised this concern, and he told me that at the time he took Mirapex, he had known all about the scientific literature linking it with ICD.

But he'd rejected the idea that he was out of control as absurd. It was only through pressure from his wife that he came to admit his problem. Fortunately, the solution is easy, but at the cost of a reasonably effective drug. Cease taking the dopamine agonists, and the compulsion soon goes away.

There was some talk about other symptoms such as depression and apathy during the meeting that day. A younger man reported sleep issues: "I used to sleep ten hours a night. Now I'm lucky if I get five or six hours . . . I just can't sleep." But, at least at this meeting, such concerns were eclipsed by an absorbing discussion about "coming out." Mario, a gentleman in his seventies, posed the question: "So far, I didn't tell anyone about my condition. Should I?" Another man who has had Parkinson's for three years admitted, "I haven't told hardly anyone . . . number one it may affect my employment, so I haven't told anyone there. In my home, my mother doesn't even know." The consensus of the group, however, was that we should not keep our disease secret. Sharon said, "It's part of who you are. You can't change it. You have to accept it and move on." And it can be helpful to reveal your disease. Bob said, "I've been in places where I've had difficulty putting my jacket on. And I'm now comfortable asking for help." Linda warned that if you kept matters secret, people might interpret your behavior—the slow movements and slurred speech—as something else. "They might think you'd been drinking . . . that you are an alcoholic." Greg talked about the importance of telling your employer as soon as possible: "That way you're protected by the Americans with Disabilities Act. If you don't mention it, then your boss might interpret your changes as laziness and fire you."

While I had told all of my friends and family that I had Parkinson's, I hadn't told everyone. When I interviewed Parkinson's researchers on the phone, I presented myself as a journalist and didn't mention that I was also a patient. My rationale for doing this was that I wanted the honest truth; I didn't want anyone holding back vital information for fear of disappointing or demoralizing me. But I wasn't entirely convinced of my reasoning. Maybe I did it because I could get away with it. When I met clinicians in the flesh, I told them up front that I had Parkinson's, because I knew they would recognize it immediately. On the other hand, with lab researchers, not trained in clinical diagnosis, I usually avoided mentioning it.

I guess the truth is that like many recently diagnosed people with Parkinson's I am driven by an ever-less-practical wish to blend into the world of the well at least part of the time. If nothing else, it's a confidence builder. It's easier to believe that I can do the things I have always done, from giving lectures to producing documentaries, if I believe my condition is mild. Part of the way I validate that belief is by moving among strangers without their thinking I look odd. Eventually, I know the disease and medications will so disable me that I will wear my disease like a scarlet letter at all times. But for now, I think my desire to look normal is an important part of my therapy.

In that room, however, we had no secrets; we were all part of the same tribe. We'd become different from other people, and that difference appeared permanent. Our common purpose was to help each other. And toward the end of the two-hour session, some people emphasized a positive side of our predicament. As Bonnie put it, "My life is so much better than it was before I got Parkinson's. Most people who know me say I'm a lot happier. I'm closer to my friends. I'm not happy I got the disease, but if I had to get a disease, I'd like this one. Unlike cancer, we're all here for a long time."

I have no idea how long my L-dopa honeymoon period will last. At the time of this writing, I have so far not encountered serious motor complications. Some patients develop fluctuations and dyskinesias within a year of starting on L-dopa; others go a decade or more without getting them. Either way, like most people with Parkinson's, I expect to take this drug in one form or another until the day I die. And that may be a long time. And because we can expect to live longer than patients in the pre-levodopa era, we all have a stake in the quest to unravel and cure our disease.

Since James Parkinson's 1817 essay, scientists had made enormous progress in establishing two of the essential foundations of this new disease. They could diagnose it clinically and also characterize the underlying pathology in the tissues. But what of the third foundational pillar of a well-understood disease: its cause or etiology? Why had Parkinson's afflicted the people in the support group but not our friends and work colleagues? Is this disease genetic in origin and passed down from generation to generation? Or is it caused by something in the environment such as a virus or a chemical toxin? Understanding the cause might well lead directly to a cure.

3

THE CASE OF THE
FROZEN ADDICTS

The nineteenth-century pioneers who struggled to unravel Parkinson's disease recognized that it did not appear to be a "familial" affliction like, say, Huntington's disease. This debilitating neurodegenerative disease, described by George Huntington in 1872, is passed from generation to generation with an "autosomal dominant" form of inheritance, in which half of all descendants (male and female) on average get the bad gene at birth and later inevitably develop the disease. Symptoms typically begin in the late thirties and progress to include loss of muscle coordination, major cognitive decline, and dementia. Huntington's is a pure genetic disease. If you have the bad gene, you will fall ill. By contrast, most people who develop Parkinson's do not appear to have inherited it in a classic dominant Mendelian fashion. No one·else in my family, for example, has ever contracted the condition.

In addition, twentieth-century researchers had investigated this question systematically by studying large registries of identical and fraternal twins. If Parkinson's is inherited, you would expect to find more cases of "concordance"—that's where both twins have the disease—among identical twins than among fraternal twins, because identical twins have identical genes and fraternal twins don't. But study after study found no difference between the two groups. Based on such evidence, Parkinson's does not seem to be primarily a genetic disease.

On the other hand, since James Parkinson's day, scientists had searched in vain for a credible environmental cause. Logically, it would have to be an exceedingly common agent—a substance found across the globe, from the largest city to the smallest village, a toxin present throughout history, from Homer's time to the present day. And this ubiquitous, persistent toxin would have to be capable of crossing the blood-brain barrier and selectively destroying the dopamine-producing brain cells in the tiny area of the substantia nigra.

From time to time, scientists have noticed that specific toxins such as manganese dust, carbon disulfide, carbon monoxide, even perhaps the cycad seed (found in Guam) cause Parkinson's-like symptoms. But on further analysis, researchers concluded that neither the clinical symptoms nor the underlying pathology truly matched the normal form of Parkinson's disease.

Genes or environment? By the early 1980s, scientists were still stuck on this fundamental question. Then a bizarre drug tragedy on the streets of San Jose shook up the world of Parkinson's research, temporarily tilting the odds in favor of an environmental cause and also giving scientists a powerful new weapon in the fight against the disease.

I have mentioned that my first skirmish with Parkinson's disease was as a journalist and not a patient. In 1985, I was living in London and working as a producer for the BBC. I got a call from *Nova*'s new executive producer, Paula Apsell, inviting me to come to WGBH Boston for six months to make a science documentary, as part of a new producer exchange program. I arrived in July, with my wife and one-year-old daughter, and started looking for stories. Very quickly, I decided that I wanted to investigate reports of some California street addicts who had mysteriously been afflicted with Parkinson's. I flew to the West Coast and spent a day speaking with the neurologist at the center of the narrative, Dr. Bill Langston. He was smart, personable, and articulate, and he told a riveting story of medical detection, a tale that any journalist would die for.*

The saga had begun three years earlier at the San Jose County Jail. On July 8, 1982, George Carillo, a forty-two-year-old drug offender, awoke to find he could neither move nor talk. He could see and hear people around him. He could smell odors in the hall. He could feel pain when he was jostled roughly by the prison guards. But he couldn't turn his head or reply if someone called his name.

Prison staff, who had watched George's condition worsen in the seven days since he'd been incarcerated, decided to transfer him to the Santa Clara Valley Medical Center emergency room for evaluation. The ER physicians thought George might be faking his symptoms to get out of jail and applied some standard "anti-malingerer" tests. They scraped his feet with the pointed end of a reflex hammer. There was no response. They applied blunt pressure to the base of his fingernails—an excruciatingly painful maneuver. But he didn't react. Finally, they placed vomit-inducing ammonium sulfate smelling salts under his nose. The mysterious inmate remained absolutely frozen.

Satisfied he was not a fake, the ER staff transferred George to the psychiatric ward, where doctors diagnosed him with catatonic schizophrenia, a rare condition in which an individual becomes mute and rigid as an emotional response to a devastating crisis. The senior psychiatrist on duty was skeptical about this diagnosis and suggested that a neurologist examine George to make sure.

*Bill Langston and I worked together on a follow-up documentary, "Brain Transplant," in 1992, and we coauthored *The Case of the Frozen Addicts* in 1995.

Neurologists came and went, but the assembled doctors couldn't agree on what was wrong. Finally, Bill Langston, the thirty-nine-year-old head of neurology, arrived at George's bedside. Langston still has vivid memories of that day. "I had never seen anything like it . . . an argument had broken out between the psychiatrists and the neurologists. The neurologists thinking that it was a psychiatric disorder, catatonic schizophrenia or something like that, and the psychiatrists thinking it was neurologic, so he had everybody buffaloed."

After examining George—by flexing his limbs and testing his reflexes—Langston became convinced his new patient had a neurological rather than a psychological problem and ordered George transferred to the hospital's neurobehavior unit. For several days, George lay there, catatonic. Then, one morning, Langston's colleague Dr. Phil Ballard noticed him moving his fingers ever so slightly. The movements were slow and looked as if they could be voluntary. On a long shot, he gently wrapped George's fingers around a pencil and slipped a yellow notepad underneath. Very slowly, the pencil started to move. After a minute, it was clear that the patient was trying to write something. Five minutes later, he had completed a name, "George Carillo." After half an hour (during which time Langston arrived at the bedside), there were five more sentences: "I'm not sure what is happening to me. I only know I can't function normally. I can't move right. I know what I want to do. It just won't come out right."

Langston and Ballard watched in astonishment. Trapped inside this frozen body was a functioning mind. Through a process of "question and answer," the clinicians started to take a medical history. Probing for clues, Langston asked George to write down any medications he was taking. Instead of listing prescription drugs, George wrote down the word "heroin."

George also told them he had a thirty-year-old girlfriend whom he'd been with just before he got sick. When they tracked down this woman, Juanita Lopez, they found she was also rigid.

"She sat like a wax doll," recalls Langston. "Her face was expressionless. Her eyes hardly blinked. She drooled continuously."

Over the next few days, Langston heard from other clinicians about four other mysterious frozen cases in the surrounding area. Two brothers, David and Bill Silvey, had been found helpless in their Watsonville apartment, unable to move or talk. And Connie Sainz, a twenty-five-year-old woman, had ended up in a mental ward in Stanford Medical Center with a diagnosis of hysterical paralysis. Connie's boyfriend, Toby Govea, who also sought help at the same medical center, appeared as rigid as a statue, while also displaying a severe parkinsonian tremor.

Langston could think of only one factor connecting all six young people—drugs. George, Juanita, David, Bill, Toby, and Connie all had a history of substance abuse. The Watsonville police had actually found several bindles of heroin in the Silvey brothers' apartment. Thinking the drugs might hold the answer, Langston procured some of the powder from the officers and sent it off for analysis. Remarkably, it turned out that the street drug was not derived from the opium poppy or from any natural product. It was typical of a new type of drug turning up on the street in the 1980s, packaged and sold like heroin but synthesized from chemicals in an underground laboratory: a so-called designer drug.

Instead of giving an opiate-like high, this concoction had turned six young people into invalids. Nurses repositioned the six patients in bed to prevent them from developing bedsores. They washed them and fed them. Despite the external appearance, their internal organs appeared to function normally. If food and water were administered in small teaspoons into the addicts' mouths, they were able to swallow it. They also seemed to have control over their bladder and bowels.

As Langston cared for his sick patients, he was struck by the remarkable similarity of their symptoms to advanced Parkinson's disease. Unlike normal Parkinson's, which generally affects the elderly and develops slowly, these individuals were relatively young and had contracted the symptoms in hours. Even though he wasn't sure it would work, Langston decided to treat them with large doses of carbidopa-levodopa. The effect was dramatic. As Langston recalls, "They came back to life. For the first time in weeks George, Juanita, David, Bill, Connie, and Toby had control of their bodies. They could move and they could talk."

Unfortunately, within days of starting L-dopa therapy, George, Juanita, David, Bill, Connie, and Toby developed severe drug-induced "motor complications," just like Cotzias's and Duvoisin's patients. "Motor complications" is a phrase that doesn't begin to describe their sufferings. As Langston says, "The cure is almost worse than the disease. All six developed one or more side effects . . . Either they were severely 'off' and frozen, or, when they turned on, yes, they could move, but they got these miserable side effects, these twisting, turning movements, and terrifying hallucinations."

But Langston realized that the drug disaster—essentially an unintended human experiment—revealed something of enormous importance. Some compound in the "heroin" had passed into the addicts' brains and, avoiding every other structure, had destroyed just that small area that makes dopamine, the substantia nigra. Identifying this toxin, he reasoned, might lead scientists to discover a related chemical in the environment that causes regular Parkinson's disease.

A vital clue came from a published report of a similar bizarre case that scientists at the NIH had investigated in the 1970s. A college student named Barry Kidston living in Bethesda, Maryland, had used a home chemistry set to make his own drugs. In the summer of 1976, after going to a library and researching the academic literature, he decided to make a compound that had originally been synthesized in 1947 by a Hoffmann–La Roche chemist, Dr. Albert Ziering. It was called 1-methyl-4-phenyl-4-propionoxy-piperidine, or MPPP. Barry realized that Ziering's formula described how to make a close analog of meperidine (better known as Demerol). Through what was in effect a designer-drug maker's trick, Ziering had made a small change in the meperidine molecule, making it some five times as potent. Barry reasoned that when injected intravenously, it would, like many painkillers, give a heroin-like high.

For several months, Barry successfully made MPPP and used it intravenously. One day in November, however, he hurried a batch, and soon after injecting it into his arm, he knew that something had gone terribly wrong. The substance burned as if it were on fire. Within three days, he froze up, became immobile, and was unable to speak. Barry was referred to the NIH,* where it was determined that he had not made pure MPPP. In addition to his target chemical, he had produced a compound called 1-methyl-4-phenyl-1,2,3,6-tetrahydropyridine, or MPTP, about which little was known.

Armed with this information, Langston and his colleagues were able to prove that MPTP was indeed the contaminant in the heroin taken by the California addicts. An underground chemist in California[†] had, like Barry Kidston, been trying to synthesize MPPP but by mistake had produced MPTP as well. And as Barry plus the six other cases made clear, this compound MPTP was a powerful neurotoxin, exquisitely suited to rapidly inducing Parkinson's.

For George, Juanita, David, Bill, Connie, and Toby, it was a monstrous tragedy. MPTP made their hard lives much more difficult. They faced a future of disabling motor complications,

---

*Barry Kidston was treated at the NIH with carbidopa-levodopa for his mysterious condition, but he continued to abuse other drugs, including codeine and cocaine. In September 1978, after eighteen months of treatment, he became very depressed. One day, he went to the NIH campus, sat down under a tree, took an overdose of cocaine, and died. An autopsy showed marked damage to his substantia nigra. The full story is in J. W. Langston and J. Palfreman, *The Case of the Frozen Addicts* (New York: Pantheon Books, 1995).
†The chemist moved to Brownsville, Texas, where he was arrested in a DEA raid in October 1984.

including violent dyskinesias and frightening hallucinations. Three of the group—George, Juanita, and Connie—later traveled to Lund, Sweden, for an experimental brain operation to graft fetal neurons to repair their brains (I'll talk more about these neural grafts in chapter 8). While the procedure helped, it didn't reverse their neurological damage. They soldiered on, growing old before their time. By 2015, all but two, Connie and Toby, had passed away.

On the other hand, MPTP was something of a godsend for Parkinson's researchers, for two reasons. First, it pointed to a possible *cause*—a chemical in the environment. One long-standing theory held that pesticides and herbicides caused Parkinson's disease. Support for this idea grew after it was discovered that MPTP was transformed in the brain into a substance with a chemical formula virtually identical to an herbicide called cyperquat.*

Second, MPTP was a powerful new research tool. It turned out that it could cause Parkinson's in monkeys as well as in human beings. The nonhuman primates would freeze up, their limbs were seized by tremors, their actions slowed. For the first time, Parkinson's researchers had an effective animal model of the disease. Rather than trying to infer what was happening from postmortem brain tissue, scientists could study Parkinson's experimentally. They could elucidate disease mechanisms and test new treatments.

Throughout history, scientists had struggled to make sense of the brain. A brain doesn't move. It makes no sound. Yet its three pounds of fatty matter perform a seemingly endless number of tasks, like walking, running, seeing, hearing, smelling, tasting, touching, loving, hating, speaking, and writing. Scientists now had the MPTP animal model to help in this complicated quest. And MPTP wasn't the only thing to come out of California in the

---

*1-methyl-4-phenylpyridinium, or MPP+.

1980s to help neuroscientists get a better handle on our most awesome organ.

The frozen addicts lived close to the part of California known as Silicon Valley, where a technological transformation was under way. Thanks to a microelectronics revolution, computers—once gigantic behemoths—had been shrunk to human size. A new industry selling personal computers had made entrepreneurs like Steve Jobs and Stephen Wozniak fabulously rich. These ubiquitous machines also provided neuroscientists with an excellent metaphor for the brain. The three-pound lump of fatty tissue could now be reconceptualized as a biological computational machine, built from neurons rather than transistors. It was a pretty good metaphor. Just as the transistors in microchips have two possible states (off and on), neurons are binary devices; they can fire or not fire. The brain's 100 billion or so neuronal switches are interconnected with two kinds of biological wires: axons that transmit outgoing messages (say, from a substantia nigra neuron) and dendrites that receive incoming ones.

The brain's hardware works somewhat differently from silicon-chip-based machines, of course. Firing neurons speak to each other when tiny packets of chemicals called neurotransmitters jump across the synaptic gap between two nerve cells. There are some 100 trillion synapses where neurons have this opportunity to communicate. A handful of neurotransmitters convey most of the messages: some neurotransmitters like glutamate tell the next neuron to get active (they're excitatory), some like GABA (which stands for gamma-aminobutyric acid) send a message to cool it (they're inhibitory), and some like dopamine can send both types of messages.

As for the brain's computer architecture, the neurons are organized in large clusters (called nuclei). These are the very structures that classical anatomists had classified according to their physical characteristics, such as the corpus striatum (striped body), globus pallidus (pale globe), and substantia nigra (black stuff). Groups of nuclei are cabled together in special circuits. "Neurons that wire together, fire together," so the maxim goes. And when they fire, these circuits carry out elaborate functions, like moving, talking, feeling, walking, and seeing. Here was a metaphor for a digital age, one that promised to link patterns of binary signals with complex human acts.

By the 1980s, many scientists were hard at work trying to figure out how such circuits "computed" in living organisms. They implanted electrodes in animals' brains and recorded neuronal firing patterns while they were awake, observing which neurons fired when the animals moved a specific body part like an arm or a leg. Perhaps the best known of these researchers is Mahlon DeLong, now in his seventies and based at Emory University in Atlanta.

DeLong, who has a gentle face and a warm, deep voice, told me he ended up working on Parkinson's disease through a lucky accident. In the late 1960s (just as L-dopa therapy was getting started), he arrived at the NIH to do a fellowship in neurophysiology, only to discover that all the hot areas of the brain to research—like the cortex and the cerebellum—had been taken by other scientists. All that was left, DeLong recalls, was a choice between the thalamus and a brain circuit known as the basal ganglia, and, he says, "I chose the basal ganglia." According to DeLong, this circuit—which links together the striped body,* the pale globe,† the black stuff (which the Russian student Tretiakoff had found was missing in Parkinson's patients' brains),‡ and a nearby structure called the subthalamic nucleus (STN)—"was a

---

*The corpus striatum, or striatum, is made up of the putamen, caudate nucleus, and nucleus accumbens.
†This is made up of an internal segment and an external segment.
‡This is made up of a substantia nigra pars compacta, which supplies dopamine, and a substantia nigra pars reticulata.

blank slate." It did appear, however, to have some connection to Parkinson's disease; for nearly a century, pathologists doing post-mortems on Parkinson's patients had noticed that this area was always damaged.

Working with healthy monkeys, DeLong set about recording the neuronal firing patterns at different places in the basal ganglia and painstakingly figured out the basal ganglia's circuit diagram.* Because this circuit was thought to be implicated in Parkinson's disease, DeLong needed a way of comparing the basal ganglia in diseased brains. Normally, monkeys don't get Parkinson's. But thanks to Bill Langston's detective work on the frozen addicts, researchers now knew that the neurotoxin MPTP would make the animals parkinsonian.

When DeLong and his colleagues recorded neuronal firing patterns throughout the MPTP parkinsonian monkeys' damaged basal ganglia, they found big differences from the healthy controls. Two key nodes in the circuit—the globus pallidus and the subthalamic nucleus—were much more active in parkinsonian monkeys.

DeLong's hypothesis was that a loss of dopamine from the substantia nigra had caused downstream nodes in the circuit to become overexcited. As a consequence, the resulting output signal from the internal part of the globus pallidus internus over-inhibits the thalamus, which in turn under-excites the motor cortex, producing the classic parkinsonian inhibition of movement.

To test the hypothesis, DeLong and his colleagues removed the subthalamic nucleus, eliminating the presumed source of the abnormal activity, to see if that would modify the monkeys'

---

*Fiendishly complicated, the circuit forms a self-contained loop. It starts with the motor cortex, connects to the input of the basal ganglia (the striatum), and splits into two parallel pathways, the "direct" and the "indirect." A combined signal exits the basal ganglia and goes to the thalamus and back to the cortex and on to the muscles. DeLong found that this motor loop was just one of five loops connecting the thalamus with different areas of the cortex, the others being the oculomotor, prefrontal, lateral orbitofrontal, and limbic loops. Decades later, neuroscientists would recognize that Parkinson's disease was far more than a movement disorder and included symptoms resulting from other loops, from eye problems to depression.

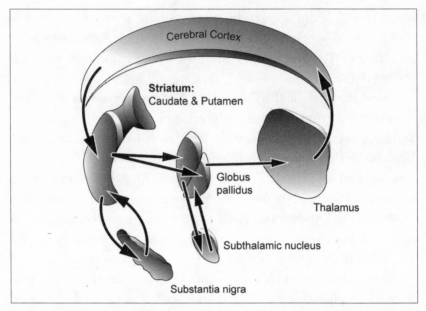

Figure 9: A circuit diagram showing a self-contained loop. The motor cortex connects to the input of the basal ganglia (the striatum) and splits into two parallel pathways, the "direct" pathway, which goes to the internal part of the globus pallidus and on to the thalamus, and the "indirect" pathway, which takes a detour to the subthalamic nucleus and external part of the globus pallidus, before continuing on to the thalamus. A combined signal passes to the thalamus and back to the cortex and on to the muscles. (Copyright © Marie Rossettie, CMI)

Parkinson's. The effect was dramatic. Says DeLong, "There was [such] an immediate reversal of slowness, rigidity, and tremor, I remember saying, 'Oh my God!' "*

While two hundred years of scientific detective work had failed to find *the* cause of Parkinson's disease, by 1990 the efforts of

*The idea of lesioning the STN was controversial. A series of human subthalamotomies performed in the 1960s had resulted in patients' getting *hemiballismus* (Latin for "half-ballistic"). Hemiballismus is highly dramatic and violent involuntary flinging movements on one side of the body. But, paradoxically, DeLong had found that removing the subthalamic nucleus in the parkinsonian monkeys had had only beneficial effects.

researchers like Langston and DeLong had convincingly traced the malady to weird goings-on in the basal ganglia deep inside the brain. This raises another question: Does this basal ganglia theory really explain what people with Parkinson's feel as they struggle to get their bodies to behave? Shortly after I was diagnosed, and with this in mind, I began asking fellow patients what it's like to have Parkinson's disease. I wanted to find out how such personal accounts from the outside squared with this picture of malfunctioning neural circuitry inside human brains.

4

MIND OVER MATTER

People with Parkinson's tell strikingly different stories about what it's like to live with their disease. While most individuals contract it in their sixties, roughly one in ten develop the condition under the age of forty. Such early-onset cases include elite athletes—such as the NBA star Brian Grant, who was diagnosed at age thirty-six, and the Major League Baseball prodigy Ben Petrick, who learned his fate at age twenty-two. And there are even reports of teenagers who develop Parkinson's. In most cases, regardless of the age of onset, the disease develops slowly, and people survive for ten to twenty years (or longer) after they've been diagnosed.

Amid such uncertainty, all of us must find the strength and wisdom to face the future with a positive attitude. Some people with Parkinson's appear to have figured out how to live long and well with their disorder, to embrace their destiny. One such remarkable individual whom I was fortunate enough to meet is the former modern dancer and choreographer Pamela Quinn.

For twenty years, Quinn performed professionally with dance companies in San Francisco and New York City. Pictures of her show a lithe, elfin performer living her dream. Then, one day in 1994, while she was reading *The New York Times*, she noticed the paper fluttering as if gently blown by a breeze. The forty-year-old soon tracked the source of the movement not to an open window but to her left hand. In retrospect, this tiny tremor was her first sign that she had Parkinson's disease. Over the next two years, Pam's symptoms got worse, spreading throughout her left arm, then to her left leg. The tremor began to affect her balance and gait. In 1996, when the Manhattan neurologist Rachel Saunders-Pullman officially diagnosed her with Parkinson's disease, Pam confronted an identity crisis. As she later wrote in an article for *Dance Magazine*, "For anyone, learning that you have a serious illness is a shock. But for a dancer, having a condition that directly affects your ability to move is profoundly shattering."

But in time, she came to realize that she could use her dancer's wisdom to help herself and others. After all, as an expert in movement she was better prepared than most to figure out the personality of a movement disorder. She knew a lot about her body: how to coax it, scold it, and trick it. "If my left hand was in tremor, I learned to calm it by shaking it vigorously. If my left foot dragged, I practiced kicking a soccer ball in a string bag to help it come forward. If my left arm didn't want to move in a full range of motion, I swung my purse from arm to arm to wake it up." Quinn might not have known it at the time, but what she was doing—this "tricking" of her body—fit with the latest neuroscientific thinking about the basal ganglia.

I went to watch Quinn teach a dance class in Sturbridge, Massachusetts, in a large hall filled with hundreds of people with Parkinson's and their caregivers. While I waited for Pam to arrive, I observed my fellow Parkies. It was hard not to imagine my future in the faces and bodies in that room. Some moved almost normally; others needed walkers to traverse the corridor. Some bent forward, shuffling with rapid, tiny steps, called festinating steps.* Some individuals in the hall twisted to the side. Many had

---

*The term "festination" is derived from the Latin word *festinare*, meaning "to hasten." In French, this is sometimes referred to as *marche à petits pas*, or "walking with little steps."

trembling limbs and masked faces. Some wobbled back and forth with levodopa-induced dyskinesias—bizarre oscillatory movements of the limbs and trunk they simply could not control.

While my condition was still mild, I was by then aware of changes in the way I moved. In fact, one of my earliest signals that all was not well, a couple of years before I was diagnosed—a sign that I initially ignored—was that I forgot to swing my arms when I walked. "Forgot" is not exactly the right term, since for most of my life I never thought about it. Indeed, few of us do. My arms had just swung automatically whenever I stepped as a natural reaction resulting from the synchronized act of walking. The right arm swings back to balance the left foot lifting into the air and swings forward again to balance the left heel landing back on the ground. The same synchrony plays out between the left arm and right leg. But suddenly, for some reason, what had been automatic now required an effort of conscious will. Then my fine-motor movements—pulling out a credit card from my wallet, cutting a steak or balancing rice on a fork, pushing icons on a touch screen without missing or undershooting—were much harder to do accurately and quickly. As Pam explained to me when I called her before going to see her class in action, these challenges came with the territory. "Those of us dealing with Parkinson's," she said, "no longer have the luxury of totally free movement or automatic functioning in daily tasks. We have to retrain our bodies."

But I had read about some striking exceptions to this impairment, such as the phenomenon of unexpected movements, sometimes called kinesia paradoxica. I'd viewed a remarkable video taken by the Dutch neurologist Bastiaan R. Bloem that demonstrates how a damaged human brain can function in some situations but not others. It starts by showing a severely impaired fifty-eight-year-old male Parkinson's patient's desperate attempts to walk along a hallway. He has trouble getting started until a clinician puts her foot across his path and he steps over it. Then, instead of walking normally, he lurches into a series of short, quick, shuffling festinating steps until he eventually falls over.

But then we see the same patient later that day, and he's outside the hospital sitting on a bike. To call it a transformation is an understatement. As if by magic, he starts riding the bike. He gracefully cycles about a hundred meters away, fluently turns his head to check for cars and pedestrians, executes a smooth 180-degree turn, and cycles back to where he started, even standing to pedal up a small incline. He looks alert, and there is no sign of his tremor. Once he dismounts the bike, he is again unable to walk, as before.

Bloem, who has devoted most of his career to investigating the gait issues faced by people with Parkinson's, has spent long hours puzzling over such cases. He notes, "My short explanation is that I just don't know. But people with Parkinson's are able to produce movements where automaticity has not been lost—this is the basis of why physiotherapy works." Patients and physical therapists do this, says Bloem, either by using alternative inputs to trigger the motor programs or by using alternative motor programs altogether. Cycling may also be a neurologically simpler task than walking. In cycling, the two legs move identically when a person pedals, whereas when a person walks, that perfect synchronization, or "time symmetry," can be lost.

I was thinking about all of this when I saw Pam walk in and sit down on a chair next to the podium. Her hands were clenched together. She seemed nervous as the moderator made the introductions. But when he announced her name over the public-address system, Pam stood up in a classic dancer's pose and commanded the room. She moved easily and displayed no sign of tremor or dyskinesia.

"I'm one of the people who's done fairly well with Parkinson's," Pam stated confidently. "I've had it now for eighteen years." She stood with perfect posture and moved confidently and fluidly before her audience. Her secret, she said, is simple: years of dance training had wired her brain and body to prepare her to wage this fight against Parkinson's. And she was here to tell us that we nondancers in the room could use this wisdom to help us move better. She listed five key insights that have helped her to manage her Parkinson's. "Dancing training involves *visual cuing*—that's using your eyes; *aural cuing*—that's using musical rhythm; *imagery*—that's knowing what quality to instill in a movement; *heightened body awareness*—you have to sense your posture in order to balance; and lastly the continuous practice of *conscious movement*—you're always telling your body what to do."

Many of her tricks, she explained, depend on music. Take my problem: walking authentically with proper arm swings. Music is usually even in tempo, so you are forced to get your four limbs working together to sync up with the beat. "When I'm outside," she told the audience, "I have my iPod and I set up a piece of music that fits my stride . . . And I step over the lines on the sidewalk—I use that as a visual cuing system to help keep my stride big. Sometimes I follow someone and copy her. The combination of the visual cuing with the music evens out my gait, gets my arms swinging, gets my legs going, and makes me happy. I consider music to be a drug."

Her advice was practical and convincing. Because people with Parkinson's tend to lead with our torsos, she says, we have to make a special effort to walk correctly and not lean into the stride so much that we trip over our feet. People with Parkinson's need to make a conscious effort when we walk. "Always feet first," says Pam. "If you're going backward, feet first. If you're going forward, feet first. If you're going to the right, right leg first. If you are going to the left, left leg first." Because we tend to scrape our feet, she told us, we need to always land on our heels. "Remember: heel, toe, heel, toe, heel, toe."

Pam cued up music with particularly rhythmic qualities—including "Girl" from the Beatles, a Hawaiian lullaby, and Peggy

Lee's rendition of "Fever." She shouted out instructions, and the people in the auditorium followed her lead. We bent our bodies, rounded our backs, stretched our spines, opened our arms, and twisted our trunks. And we smiled.

Pam offered some tips for handling advanced movement glitches. Many in the audience struggled with the bizarre but fascinating phenomenon of "freezing of gait." Advanced Parkinson's patients—like Bastiaan Bloem's bicycle patient—walk along and suddenly get stuck to the floor, as if their feet were glued down. They freeze like statues. This embarrassing breakdown typically happens when approaching small, constricted spaces—going through doorways, entering elevators, crossing busy streets, or just before turning. Equally extraordinary are the tricks that counter the freezing behavior. Neurologists know that simply drawing a chalk line on the floor will magically trigger some compensatory brain circuit and enable the frozen individual to move forward, stepping over the line; much like the clinician who put her foot in front of Bloem's patient, placing a foot in front will also bypass the mental blockage: the Parkinson's patient can simply step over the foot and then resume walking. But what if you're all alone when you freeze? Pam offered some solutions.

"You have to shift your weight sideways," she told the crowd. "When you're frozen, you are trying to go forward. Your torso's going forward . . . and your feet aren't. But if you shift your weight sideways, it takes the weight off one foot and allows it to move forward." Says Pam, "I find the image of a penguin helpful. Begin walking side to side like a penguin, and transition to a more normal forward walk."

She offered us a different image to avoid stooping while walking. "I think of fashion models. They're always leaning back—their legs forward. So when my meds start to wear off and my posture starts to change, I think: fashion model. That's my imagery. I become that person. It gets my weight back; it gets my feet forward."

Pamela Quinn is certainly an outlier, with a slowly progressing form of the disease. But we can all learn from her. Her wisdom exhorts Parkies to keep active, to mindfully circumvent gait and balance issues. As she puts it, "We must treat the mind as a muscle; it needs to be strengthened and made flexible just as much as our legs and core muscles."

As I drove home after the class, I felt inspired. Pam's demonstration had reminded me that the human brain is constructed with considerable built-in redundancy. Which means that there are often several ways to accomplish the same goals.

In addition, Pam's analysis of the challenges of moving with Parkinson's disease seems to conform with what scientists have discovered about the neurological changes going on inside our brains. In Parkinson's, as scientists like Mahlon DeLong have shown, the death of neurons in the substantia nigra (which normally sends dopamine to the striatum) causes the basal ganglia circuitry to become dysfunctional. Neuroscientists agree that this piece of hardware, while tiny, is very versatile. The basal ganglia normally acts as a kind of adviser that helps organisms learn adaptive skills by operant conditioning—rewarding good results with dopamine bursts and punishing errors by withholding the chemical. For example, when a tennis player practices a stroke over and over again, the basal ganglia circuitry both rewards and "learns" the correct sequence of activities to produce, say, a good backhand drive automatically. Human infants rely on the basal ganglia to learn how to deploy their muscles to reach, grab, babble, and crawl—motor sequences that will help them adapt to their changing world. Over time, an individual human accumulates a huge repertoire of complex tasks that she can accomplish automatically without thinking, thanks in large part to the basal ganglia.

A few days after Pam's address in Sturbridge, I spoke to Pietro

Mazzoni, head of the Motor Performance Laboratory at Columbia University Medical Center. I wanted to better understand the
physiology behind my fine-motor clumsiness.

The researchers in the lab study how the brain controls limb
movement in both healthy individuals and those with neurodegenerative diseases. In particular, the forty-five-year-old
Mazzoni investigates the basal ganglia by comparing the per-
formances of healthy individuals and Parkinson's patients
on everyday motor tasks from walking to grabbing a hairbrush.

"The motor system is extraordinarily sophisticated," says
Mazzoni. "Imagine a simple act of reaching out your right arm
to pick up an object." To arrive at the target, he says, some muscles need to be contracted, others extended. You need to decide
how fast and how far you are going to move, and you need to
open your hand and grasp the object with the right amount of
force. Each element in the sequence must be timed precisely. But
Mazzoni says it's more complicated than this. "It's all about the
context. It's not enough to judge speed and distance. You also
need to decide how rigidly to hold your arm. And that depends
on what your goal is. If you're picking up a valuable glass object
that's balanced precariously on the edge of a shelf, or if your
target is a hot cup of coffee filled to the brim, then you will want
your arm to be more rigid. If, on the other hand, you are trying
to scoop up a bouncing ball, where you need to switch directions
quickly, then you need to be loose, not rigid." The ability to pull
off such complex sequences of steps automatically, without thinking (or at least thinking that you're aware of), depends on the
basal ganglia. Likewise, the basal ganglia provides you with the
ability to multitask and, say, sit and eat spaghetti while tapping
your feet in time to music and listening to a family member talk
about her day.

But this brain circuit has a vulnerability: the basal ganglia
depends on dopamine. Without dopamine, the signals passing

through the basal ganglia get garbled, and it ends up giving "poor advice" to the cortex. This is one of the reasons why Parkies have trouble picking up small objects and moving around fluently: the motions are too hesitant, too small, too slow, too rigid, too shaky, too feeble, and badly timed. These are symptoms of a brain in conflict with itself.

Having Parkinson's feels a bit like going on vacation in another country and having to drive on the "wrong" side of the road. Driving is one of those activities that we outsource, in large part, to the basal ganglia. The American driver's basal ganglia has picked up the habitual behavior of driving on the right-hand side of the road through thousands of hours of practice navigating a car through the streets of the United States. When he tries driving in England, these learned habits are a liability. To compensate, the American motorist must invoke the conscious, deliberate, mindful, and goal-directed part of his brain—the cortex—to override the basal ganglia. The driving will be difficult, partly because the conscious brain is now doing all the work, but mainly because it's having to compensate for the basal ganglia's signals that are inappropriate for the situation at hand.

Deprived of adequate supplies of dopamine in the brain of someone with Parkinson's, the circuits of the basal ganglia misbehave. Now corrupted signals pass to other brain regions such as the thalamus (which relays sensory and motor data to the cerebral cortex) and the cortex itself (which is responsible for many higher functions such as thinking, language, and consciousness). These bad signals disrupt communication between the brain and the muscles, resulting in a series of classic symptoms: tremors, slowness, weakness, rigidity, stooped posture, the tendency to walk without arm swings, shrinking handwriting, low voice volume, and so on. Like the conflicted American driving in England, a patient has to use conscious, mindful, deliberate, and goal-directed thoughts to override the basal ganglia, willing the body to stand up straight, to swing the arms rhythmically when walking, to lift the feet so you land on your heels, to write clearly, to speak as loudly as possible.

Pam Quinn's tricks seemed to work along just these lines. She had effectively bypassed the basal ganglia, using other parts of her brain—mainly her cortex—to pull off consciously what a healthy basal ganglia accomplished subconsciously. She'd used visual and auditory cues and conscious imagery. She'd broken down sequential tasks into individual steps and more. So while it's tragic that an advanced Parkinson's case, for example, may walk with smaller and smaller steps and eventually freeze as if stuck to the floor, it's somewhat gratifying that the simple remedy of moving side to side may work around the storm in this part of his brain.

By combining the scientific insights of researchers studying neurochemistry and neurophysiology with the personal experiences of patients like Pam, I was beginning to get a deeper understanding of what had befallen me and the challenges I would face down the road. But like other people with Parkinson's disease, I wasn't just interested in understanding what was wrong with me. A part of me hoped for a cure.

5

PATIENT POWER

There appears to be sufficient reason for hoping that some remedial process may ere long be discovered, by which, at least, the progress of the disease may be stopped.

—James Parkinson

People react to the news that they have a serious disease like Parkinson's in different ways. Some get demoralized and withdraw from the world, as I did when I first received a diagnosis. Some, like Pam, decide to help fellow patients to manage their symptoms. A few emerge as transformative figures in disease advocacy. Such individuals become leaders in the scientific quest to defeat Parkinson's disease itself, ready, if necessary, to disrupt the medical research system in order to speed progress to a cure.

The same year that Mahlon DeLong reported his study of parkinsonian monkeys, a young actor received some disturbing news. In November 1990, Michael J. Fox, the beloved star of the *Back to the Future* trilogy, was in Gainesville, Florida, filming the comedy *Doc Hollywood*, when he noticed his left pinkie was trembling. The twenty-nine-year-old Fox consulted a local neurologist who examined the finger and said he was fine. "In his opinion," Fox recalls, "the source of the finger spasms was most likely a minor injury to my ulna. 'You mean my funny bone?' With a confirming nod, the doc joked that wasn't it appropriate, given what I did for a living. We had a nice little chuckle over that one."

It wasn't funny. The shaking didn't go away. The tremor spread through Fox's left hand, and his shoulder became stiff and painful. Then, in August 1991, during a family vacation in Martha's Vineyard, Fox decided to go for a run. The young actor, who had displayed vigorous athleticism in *Back to the Future*, struggled to keep going. His wife, Tracy, noticed immediately: " 'You look like hell,' she said. 'The left side of your body is barely moving. Your arm isn't swinging at all. I don't think you should run anymore until you get a chance to see a doctor.' "

In October of that year, a New York City neurologist diagnosed Fox with early-onset Parkinson's. Fox responded the way many recently diagnosed Parkinson's patients respond: with denial and secrecy. He sought second opinions, but all the experts agreed: he had Parkinson's disease. So, avoiding neurologists altogether, Fox persuaded his internist to prescribe medications, including levodopa. Says Fox, "Therapeutic value, treatment, even comfort—none of these was the reason I took these pills. There was only one reason: to hide. No one, outside of family and the very closest of friends and associates, could know. And that is how matters stood for seven years."

In 1998, Fox was working on his third season of the sitcom *Spin City*, and still neither his fellow cast members nor the audience had guessed his medical secret. As he tells it in his book *Lucky Man*, "I can vividly remember all those nights when the studio audience, unknowingly, had to wait for my symptoms to subside. I'd be backstage, lying on my dressing room rug, twisting and rolling around, trying to cajole my neuroreceptors into accepting and processing the L-dopa I had so graciously provided."

Fox's seven-year period of deception and denial depended on his actor's bodily intuition, levodopa, and timing. When Fox sat in his dressing room and swallowed a tablet, he knew that he was initiating a complex process. The carbidopa-levodopa tablet passed into his stomach and traveled to his upper gastrointestinal tract, where it was absorbed into his bloodstream. As for all patients who take levodopa, over 90 percent of the drug was chemically broken down and never made it to Fox's brain. But some arrived there, was selectively taken up by his remaining dopamine nerve terminals, and converted into dopamine. Almost immediately, his rigidity would melt away, his tremor would largely disappear, and he could move more fluently. He got back a lot of other abilities as well, from facial expression to blinking, from swallowing to swinging his arms when walking.

But this process took time, perhaps an hour or more before the full effects kicked in. Fox became a master of timing his medications so that he would be on the set when the levodopa had reached an optimal level. But any production delay put his plan in jeopardy. For inside his brain, the levels of levodopa peaked and then fell according to the laws of biochemistry, and the parkinsonian symptoms started to return. Three and a half hours after popping the tablet, Fox would be pretty much back where he started. To conceal his annoying tremor, he used some classic tricks: sitting on one hand, holding on to furniture, or keeping one or both hands in his pocket.

In 1998, in an attempt to eliminate his left-side tremor once and for all, Fox underwent a thalamotomy operation in Methuen, Massachusetts. This relatively uncommon surgical procedure had been developed in the 1950s, before there were any effective anti-Parkinson's medications. Under top secret conditions, Dr. Bruce Cook, a Boston neurosurgeon, surgically destroyed part of Fox's thalamus on the right side of his brain, a region that regulates involuntary movements on the left side of the body. Cook operated on Sunday morning, when the hospital was virtually empty. Fox hoped this surgical fix would enable him to continue his ruse. As he put it, because "my symptoms were still limited to my left, less dominant side (I'm right-handed), a cessation of tremor there would be as good as a return to normal."

The operation worked. The tremor in his left side disappeared immediately after the surgery. But the shaking soon reappeared, this time on the right side of Fox's body. As Fox puts it, "I've known for years that this was an inevitability. I have Parkinson's disease; it's a progressive disorder. It's just doing what it's supposed to do. What was I supposed to do now?"

By this point, his frequent trips to Boston for medical consultations had not gone unnoticed. The circling media began to close in. Entertainment reporters started using the "Parkinson's" word in their questions. Fox decided after nearly a decade of denial it was time to come out. He agreed to two interviews, one with Barbara Walters of *20/20*, the other with Karen S. Schneider and Todd Gold of *People* magazine.

In the *People* magazine interview, Fox talked about what he'd learned from the experience. "It's made me stronger. A million times wiser. And more compassionate. I've realized I'm vulnerable . . . no matter how many awards I'm given." In his interview with Barbara Walters, Fox displayed irrepressible optimism that he still might be saved from his Parkinson's: "There are so many things on the horizon. So many medications, and surgical procedures . . . I really feel that within the next years they're going to find a way to flick a switch, and this is gone." Walters asked him if he thought there would be a cure by the time he reached fifty years of age. Fox, thirty-seven years old at the time, seemed in no doubt. He told Walters, "I know I won't have this. I will not have it."

Fox's story generated enormous public interest not just in his personal situation but also in his disease. As he notes, "Without intending to, I had sparked a national conversation about Parkinson's disease." He began to see that contracting this terrible malady was not all bad. A lot had been stolen from him, but he had received a gift as well, a chance to play a very different role than he would have done as an actor, an opportunity to join the fight to cure a disease that affects all of us—because the more we know about the Parkinson's brain, the more we know about this master organ's functions and failures.

Meanwhile, across the Atlantic, another drama was in play. In 1996, on November 4, the eve of Guy Fawkes Day—the day when the British build bonfires and set off fireworks, in remembrance of Fawkes's unsuccessful 1605 plot to blow up the Houses of Parliament—a twenty-seven-year-old London-based surveyor named Tom Isaacs was diagnosed with Parkinson's disease. Like others before him, he struggled to come to terms with his new identity. As he put it, "The truth was that I was now trapped inside a body that was now aging at an alarming rate and was often incapable of responding to the demands put on it."

Then, one night, Isaacs had a dream that would change his life. It was a dream about a group of highway workers digging holes in the road with pneumatic drills. "In the dream," Isaacs recalls, "there was a power cut, and the workers laid down their tools and went off to get a cup of tea. I walked into the cordoned-off area, picked up a drill, and with my tremor I was able to [drill] a big hole." Isaacs says he woke up with a smile on his face thinking that maybe he could make something positive out of his Parkinson's.

Isaacs conceived of a plan to raise money for Parkinson's disease—a sponsored walk. After warming up with a charity walk from one end of Britain (John o' Groat's in Scotland) to the other (Land's End in Cornwall), he came up with a grand vision to raise money for Parkinson's research—to hoof it around the entire coastline of Great Britain, a distance of forty-five hundred miles.

He sought support from Sir David Jones, the chief executive of the retail empire NEXT PLC, who also had Parkinson's. When Jones heard about the ambitious scheme, Isaacs recalls him saying,

"You're a complete nutter," but then Jones wrote him a check for £20,000. Isaacs's bold plan was up and running.

On April 11, 2002, Isaacs set off on his expedition. Over the next 365 days, moving in a counterclockwise direction round Great Britain, he averaged fifteen miles a day, wore out five pairs of boots, lodged at 238 different bed-and-breakfasts, and raised £350,000 for Parkinson's research. One thousand thirty-two people joined him for parts of the journey. On April 11, 2003, after a grueling trip in which he pushed himself to the limit (chronicled in his memoir *Shake Well Before Use: A Walk Around Britain's Coastline*), he arrived back in London, where he was greeted by his fellow Parkie Sir Richard Nichols, former lord mayor of London. As if this weren't enough, two days later Isaacs completed the London Marathon in approximately five hours.

Soon afterward, Isaacs, Jones, and Nichols—together with Nichols's close friend the air vice-marshal Michael Dicken (who also had Parkinson's)—met up for dinner. The four men agreed that simply giving the £350,000 to the Parkinson's Disease Society (later called Parkinson's UK) would be a mistake. Despite an admirable record of supporting patient care, this charity had little history of funding research into a cure. Back then, Isaacs recalls, "the word 'cure' was never used . . . you know it was forbidden. Cure was seen as a false hope. And actually, if you don't have hope in Parkinson's disease, you don't have anything."

As Isaacs says, the consensus was "that we needed something focused on the research not just on the care . . . and it needed to be a bit more edgy, a bit more feisty." Because no such organization existed, they decided to create one. In 2005, Isaacs, Jones, Nichols, and Dicken founded Cure Parkinson's Trust, a charity dedicated to finding a cure for the malady that had affected all of them, as well as millions of others worldwide.

·

Back in the United States, Michael J. Fox had been busy trying to decide what to do with the rest of his life. The leaders of the major Parkinson's charities based in the United States realized that his celebrity could bring sustained attention to his disease. As Fox put it, "By the end of 1998, my desk was covered with correspondence bearing the letterhead of various Parkinson's organizations across the country. All of them wanted my help." Just as Magic Johnson put a human face on HIV/AIDS, Fox, they speculated, might come to personify Parkinson's.

After meeting with many of the foundations, Fox came away somewhat disillusioned, feeling that many of these groups viewed themselves as competitors uninterested in working together. One Parkinson's activist who impressed Fox was Joan Samuelson of the Parkinson's Action Network (PAN). Samuelson, like Fox, had early-onset Parkinson's disease. A California lawyer, she had been diagnosed in 1987 at age thirty-seven. Shortly afterward, she resolved to devote the rest of her life to Parkinson's advocacy. As she recalls with a chuckle, "I just decided that was the way I was going to get cured—by political action." She resigned from her law firm, founded PAN, and became its first president.

It didn't take Samuelson long to realize that Parkinson's disease was not getting a fair share of federal research dollars, so she threw herself into political activism. "I was told by a luminary in the NIMH [National Institute of Mental Health] that Parkinson's wasn't receiving any money because it wouldn't accomplish anything . . . Just ludicrous . . . [It was] just an excuse." On a plane ride back from Washington, D.C., where she had been meeting with the California congressman Henry Waxman, Samuelson reviewed "the disease list." This was an NIH publication that listed every disease and the amount of money the NIH was spending on it. "And it was horrifying. Parkinson's was getting about $25 million, period, which is nothing in comparison with hundreds of millions of other diseases . . . and already I could see that the money drove research activity. So we had to have more money."

In September 1999, partly as a result of her efforts, Congress scheduled a Senate hearing on Parkinson's research, and Samuelson asked Fox if he would join her and testify. The goal of this hearing was to focus attention and urgency on the quest for a cure.

Senator Arlen Specter of Pennsylvania brought the meeting to order. "We have a hearing today which focuses on Parkinson's disease. This is a medical problem of enormous impact. We have with us today a very distinguished panel, including Mr. Michael J. Fox."

The hearing—which would solicit testimony from scientists, advocates, bureaucrats, and patients—was packed. Specter, in a bullish mood that day, remarked on several occasions that researchers were within five years of conquering Parkinson's. He addressed Gerald Fischbach, the recently appointed director of the National Institute of Neurological Disorders and Stroke (NINDS), with these words: "The subcommittee has heard testimony . . . that it is realistic to conquer Parkinson's within five years. Now that was about a year ago, so I guess it is four years now. A two-part question. Would you concur that we are that close to solving Parkinson's? Second, what could we do to even make it a shorter time interval to conquer Parkinson's?"

Gerald Fischbach seemed ready to go along with Specter's optimistic stance.

"I concur that we are close to solving—and I mean the word 'solving'—Parkinson's disease. I hesitate to put an actual year number on it. I think, with all the intensive effort, with a little bit of skill and luck, five to ten years is not unrealistic. We will do everything possible to reduce that below five years. I would not rule that out."

Specter asked, "Will more money enable you to do it in less than five years?"

Fox was captivated by what he heard. "If one of the congressional cameras had me in an isolated close-up, I'm sure I executed one of the finest double-takes of my career. I'd expected Fischbach to express confidence and lay down a challenge for researchers to match it and for Congress to support them in any way possible, but I wasn't expecting him to suggest a timeline. His testimony energized me. This was doable. I had been given the idea that a cure was possible, and I needed to act upon it."

While a five-year timeline seemed far-fetched, it's fair to say that in 1999 Parkinson's research looked very promising. The neuroscientist Bill Langston told the senators, "I think most of us feel that this disease can be solved. It may be the first of the (neurodegenerative) diseases to be solved. But we must pursue every lead relentlessly if we are going to get there."

In 1999, there were two main strategies for helping patients. First, there were scientists working on taming the symptoms of the disease: everything from better drugs to exercise to surgery. A second category of research programs sought to modify the underlying disease: to slow, stop, and perhaps reverse it. Many scientists were intrigued by the concept of neuroprotection; researchers were testing a class of drugs known as monoamine oxidase inhibitors, like selegiline, to see if they protected the neurons from environmental toxins and thereby slowed the progression of the disease. Another pharmacological strategy was to try to rescue damaged neurons using a kind of brain nutrient known as growth factor. Such growth factor could be infused directly into the brain or indirectly via gene therapy—a technique where viruses were used to carry growth factor genes into the brain. Said Fischbach, "If we can turn on even half of these remaining brain cells, we might be able to reverse the parkinsonism entirely, and there are substances that may do this."

Also discussed at the meeting was the controversial idea of

repairing brains by grafting new dopamine-making nerve cells dissected from fetal tissue. For more than a decade, scientists in Europe and the United States had grafted such fetal dopamine neurons—dissected from aborted fetuses—into the brains of Parkinson's patients. Three of the frozen addicts had traveled to Sweden for this experimental procedure. There was hard evidence that many of the grafts survived and flourished, making new neural connections and pumping out dopamine. And there were compelling data that in some mostly younger patients, at least, the grafts reversed the crippling motor symptoms of Parkinson's disease, enabling them to manage on much lower doses of levodopa. Equally controversial were the prospects of brain repair offered by embryonic stem cells, miracle entities that could morph into any kind of cell: heart, kidney, liver, and dopamine neuron. Fischbach said this technology offered "a potentially unlimited supply of dopamine cells."

By the time it was Fox's turn to address the committee, he was convinced that this was, as he put it, a winnable war. "I did not graduate from high school," he told Specter, "but I learned enough Latin to be able to say this—*carpe diem* . . . We are there with this. We are really there. If we can just get a focus on it, I really think we can get this done." At that moment, Fox realized he might use his celebrity to increase attention and support for his disease. He resolved to quit acting, at least for a while, and devote himself to finding a cure for Parkinson's. Rather than lend his name to an existing Parkinson's organization, he would create his own. "I realized that I was ridiculously unqualified to contribute to this effort in any substantive way; I wasn't an MBA or a PhD . . . But my optimism had crystallized into definitive hope."

Over the next few months, Fox designed his own foundation dedicated to Parkinson's research. His goal was to energize the patient community, raise large sums of money, identify underfunded scientists with good ideas, and provide the support they needed, all as quickly as possible.

In May 2000, he officially launched the Michael J. Fox Foundation for Parkinson's Research, appointing Debi Brooks—former VP in the fixed income and asset management division at Goldman Sachs—executive director. During her job interview, Fox had explained that her position was only temporary. They would not build an endowment and live off the dividends like other foundations. Fox's plan was to spend money as soon as they raised it. More important, after a cure was found, there would be no reason to continue. As Fox told Brooks, "The last thing I want is for you and I to find ourselves discussing our twentieth annual fund-raiser. In fact, if that day ever comes, you're fired." Fox's pitch to advisers was the same: "I need you to help me go out of business . . . If we can find a cure for Parkinson's, our work is done." The business plan was to fund the foundation's work through donations. Proceeds of Fox's first book, *Lucky Man*, and initial gifts generated a $1 million starter fund. The foundation told researchers that applications from investigators were due February 1, 2001, and that the submissions would be reviewed (by a scientific board put together and chaired by Bill Langston) in just six weeks. By comparison, the NIH typically spends a year reviewing applications before issuing grants. There was skepticism that scientists could act that nimbly. But the response astonished everybody. By the six-week deadline, the new foundation had received 220 grant applications from twenty countries.

Energized by the Michael J. Fox Foundation in the United States, and to a smaller extent by the Cure Parkinson's Trust in the United Kingdom, scientists around the world doubled down on research with newfound enthusiasm and optimism. Nearly two centuries after James Parkinson published his essay, researchers were hard at work, seeking not only ways to treat the symptoms but also radical disease-modifying solutions—ways to protect, revive, or replace sick neurons. If scientists were going to put Fox's foundation out of business before his fiftieth birthday, there was no time to waste. In the meantime, Parkies depended on symptomatic therapies, interventions that helped them make the best of a bad situation. These included drugs like L-dopa, exercise, rehabilitation therapies, and a remarkable new surgical procedure that had been discovered by accident.

6

SURGICAL SERENDIPITY

Deep brain stimulation is a bit like real estate. Success depends on location, location, location.

—J. William Langston, MD

The thalamotomy that Michael J. Fox underwent in Methuen, Massachusetts, emerged from a heroic neuro-
surgical tradition that peaked in the first half of the
twentieth century. Medical historians generally use the label "heroic" to refer to bold procedures of last resort that risked the lives
of patients and the reputations of their surgeons. Interest among
neurosurgeons began with the observation that Parkinson's patients who suffered small strokes—and experienced, for example,
numbness in the face, arms, or legs on the opposite (contralat-
eral) side of the body—sometimes also experienced a relief in
their parkinsonian symptoms such as rigidity and tremor. This
observation inspired some neurosurgeons to take bold and arguably reckless psychosurgical forays into the brain to see if directly lesioning certain areas brought patients similar relief.

Initially, the basal ganglia was not one of their targets. The
influential American neurosurgeon Walter Dandy—famous for
his speed and dexterity (he reportedly performed a thousand

operations a year in his prime during the first half of the twentieth century)—argued that the basal ganglia was the seat of
consciousness and should be left alone, proclaiming *noli me
tangere*, "let no one touch me." But attitudes changed after the
surgeon Russell Meyers safely cut off the head of a Parkinson's
patient's caudate nucleus (a part of the basal ganglia) in an effort to relieve the symptoms.

Then, in 1952, a New York neurosurgeon, Irving S. Cooper,
made a serendipitous discovery. He was removing a bundle of
nerve fibers in the midbrain of a Parkinson's patient as part of a
procedure called a pedunculotomy, which some neurosurgeons
of the day thought could modestly mitigate parkinsonian symptoms. Midway through the operation, Cooper's hand slipped, and
he severed the patient's anterior choroidal artery. The surgeon was
forced to ligate the artery and terminate the botched operation.
But astonishingly, after the patient awoke from surgery, even though
he had not received a pedunculotomy, his tremor and rigidity were
gone, and his motor and sensory faculties were still intact. Because
this critical artery supplies blood to the basal ganglia, Cooper and
other surgeons reasoned that cutting out or ablating those brain
nuclei directly might also effectively relieve the symptoms of
Parkinson's disease.

And so it often did. Surgeons referred to one such procedure
as a pallidotomy, in which they excised the globus pallidus internus (a key part of the basal ganglia), and it could mitigate all the
cardinal symptoms of Parkinson's—tremor, rigidity, and bradykinesia (the slow and impoverished movements). In a different
operation, surgeons found that removing part of the thalamus, a
brain center that relays sensory and motor data to the cerebral
cortex, relieved just the patients' tremor symptoms. This was the
forerunner of the thalamotomy that Fox would undergo in 1998.

Even though there was no convincing theory to explain the
symptomatic improvements of Parkinson's patients, neurosur-

geons performed hundreds of pallidotomies and thalamotomies across the globe—until, that is, the dawn of widespread L-dopa therapy in 1971. As the Toronto neurosurgeon Andres Lozano told me, "After L-dopa, we went into what I call the Ice Age of surgery for Parkinson's; surgery just about vanished."

But by the late 1980s, as increasingly large numbers of patients around the world began to confront serious and intractable L-dopa-induced motor complications—the involuntary movements and sudden gait freezing—some surgeons started to offer surgical options to patients again. Lauri Laitinen, a Finnish surgeon working in Sweden, had also devised a safer and more effective way of performing pallidotomies. Meanwhile, Mahlon DeLong's classic experiments from the 1980s, in which he revived monkeys rendered parkinsonian by the neurotoxin MPTP (found in the California street drug), had created a stir among Parkinson's researchers. DeLong had simply cut out parts of the parkinsonian monkeys' basal ganglia—removing the overexcited subthalamic nucleus—and the animals could move again, and their functioning didn't appear to be compromised as a result of losing this brain center. This astonishing finding contributed to a new scientific understanding of what neurosurgeons had previously discovered by trial and error and provided a compelling rationale for reviving Parkinson's disease neurosurgery in the basal ganglia.

In 1990, Mahlon DeLong, along with the surgeon Roy Bakay and the neurologist Jerry Vitek, started a pallidotomy program at Emory University. They performed their first pallidotomy on a human in 1992. DeLong says it was "extremely, remarkably successful . . . even better than the MPTP monkeys." It not only improved rigidity and bradykinesia but also greatly lessened tremor and dyskinesia. Scientifically, the surgery's success was extraordinarily interesting. Apparently, patients were better off living with no basal ganglia than with a dysfunctional one. To this day, the operation's efficacy remains a puzzle. As DeLong says, "How can we explain the fact that you can completely interrupt the output from the motor circuitry of the basal ganglia and restore movement and function?"

Even without a good explanation, people struggling with Parkinson's flocked to Emory and other similar centers for pallidotomy operations. But there were downsides. Neurosurgery—no matter what type—carries inescapable risks of hemorrhage, stroke, and infection. To limit these risks, most neurosurgeons were willing to offer Parkinson's operations only on one side of the brain. According to Andres Lozano, operating on both sides would be just too dangerous. "If you only need to lesion one side of the brain you can get by . . . but if you need both sides done, then you run an unacceptably high risk of complications." Even if the surgeries are done at different times, the additive risk is significant. It followed that patients with Parkinson's symptoms on both sides of their bodies were out of luck.

In time, neurologists noticed some other limitations with pallidotomies. The University of Michigan neuroscientist Dan Leventhal says the procedure likely had consequences for the future performance of the basal ganglia. "Pallidotomy patients moved pretty well, but when it comes to learning a new sequence of activity like something on the piano, they really have a hard time doing it. They lose the ability to learn new things."

Pallidotomies are also irreversible. Once a part of the brain is destroyed, it can't be restored. How much better for patients it would be if the benefits of surgery could be won without excising neural tissue. Remarkably, a French neurosurgeon stumbled on a way of doing just that.

In 1987, Alim-Louis Benabid of the Joseph Fourier University in Grenoble, France, made a striking discovery. The way the elegant seventy-year-old Benabid tells it, his epiphany happened during a routine thalamotomy operation. He was planning to vaporize part of a Parkinson's patient's thalamus (and eliminate his tremor) using a high-frequency electrode heated to around two hundred degrees Fahrenheit. Like all good neurosurgeons, before starting this destructive and permanent procedure, Benabid did a routine check to make sure he'd placed his electrode in the optimal location. Typically, this was done by firing up the electrode and asking the patient if he felt anything. If the electrode is positioned too far behind the target region, the patient feels a tingling sensation. If it's placed too much to one side of the target, the patient will feel contractions in his hand or in his face.

But Dr. Alim-Louis Benabid was curious. What would happen, he asked himself, if he stimulated the site—a part of the thalamus called the Vim—with different electrode frequencies? So, proceeding in small increments, he tested the effects of different frequencies up to 130 hertz. As he later remarked, "This is how I got lucky." Lucky indeed. For at 130 hertz, the device's maximum frequency, the patient's tremor simply disappeared. While low frequencies of 5–10 hertz actually caused tremor, high frequencies did the opposite and mimicked ablation. "The easiest explanation," he said, "was that stimulation, above a certain frequency, disrupted, distorted, or altered the neuronal message."

Benabid—who, unusually for a neurosurgeon, has a PhD in physics in addition to his medical qualifications—went on to collaborate with Medtronic, the U.S. manufacturer of cardiac pacemakers, to produce a practical implantable electronic device to treat essential tremor and prove its safety and efficacy to regulatory bodies such as the FDA. The apparatus consisted of two thin wires—the stimulating electrodes—placed near the thalamus in each hemisphere, connected to a small pacemaker-sized battery, positioned in the patient's chest. Once the device was activated, the disabling tremor was eliminated. No brain tissue needed to be cut out. In Benabid's words, this operation could be thought of as a "reversible lesion." By 1993, Canada, Europe, and Australia had approved the Medtronic device for essential tremor, and the United States followed suit in 1997.

But why stop with the thalamus? What about applying the technique to other brain centers that regulated movement, such as the globus pallidus and the subthalamic nucleus? In 1993, inspired by DeLong's work lesioning parkinsonian monkeys' subthalamic nuclei and reversing their symptoms, Benabid tried *stimulating* these brain nuclei in the basal ganglia rather than cutting them out. It worked well in monkeys, so he tried it in human beings. Again, it succeeded, radically reducing rigidity, tremor, and bradykinesia. Following the procedure, he was able to reduce his patients' L-dopa medication by half. Disabling dyskinesias, the result of excessive dopaminergic medication, often vanished. In 1994, the Swiss surgeon Jean Siegfried performed an operation stimulating another key part of the basal ganglia, the globus pallidus internus—the structure excised in a pallidotomy—with similar benefits.

Benabid and Siegfried had convincingly proved that it wasn't necessary to destroy parts of the basal ganglia in order to shut off its corrupted signals. Electrical stimulation could achieve the same goals with less risk, making thalamotomy operations for essential tremor and pallidotomy operations for Parkinson's disease effectively obsolete.

Surgeons, researchers, and scientists began to refer to this new electrical procedure as deep brain stimulation, or DBS. While profound questions remained about just why blocking the output of a critical brain circuit restored function, it was potentially the biggest therapeutic advance in Parkinson's research since L-dopa. By 2003, the FDA had approved the DBS procedure for Parkinson's disease as well as for essential tremor. The approval permitted neurosurgeons to implant stimulators into two sites in the basal ganglia (either the subthalamic nucleus or the globus pallidus internus).

Among the early recipients of DBS, ironically, was Toby Govea, one of the frozen addicts. In 2003, Toby was incarcerated in Atascadero State Hospital, a maximum-security hospital in the California prison system, and was in bad shape. Heavily medicated (L-dopa seven times a day, plus Mirapex, clozapine, and more), Toby had tremors and dyskinesias 50 percent of his waking hours and was "off" 25 percent of the day. Additionally, he suffered frequent hallucinations and pervasive gait freezing, which often led to falls. When offered the chance to undergo this new procedure, Toby consented, and on October 28, 2003, he was prepped for surgery at the University of California San Francisco Medical Center. The neurosurgeon Phil Starr (who had trained with Mahlon DeLong at Emory University) inserted stimulating electrodes in both sides of Toby's brain, with the hopes of ameliorating some of his motor symptoms. The operation was successful. With the stimulators adjusted, Toby's doctors could cut his medications by half. His dyskinesias and hallucinations vanished, and his amount of "on" time improved significantly.

Deep brain stimulation came with all the usual risks of brain surgery. It wasn't, therefore, generally recommended for patients over seventy or for individuals with cognitive problems such as dementia. And doctors had to remind patients that it wasn't a cure. The therapeutic benefit only continued as long as the electrical stimulation persisted. Switch off the stimulator, and the symptoms immediately returned. When the battery ran out of power after a few years, the patients' symptoms would reemerge, until a fresh power pack was surgically installed. In addition, deep brain stimulation didn't halt the disease's progress. Even as some motor problems lessened, new issues, often related to balance, could emerge. Despite such caveats, over the next decade some 100,000 people worldwide elected to undergo deep brain stimulation. One of them was Nancy Egan.

·

Nancy, a slender seventy-six-year-old woman from Norwood, Massachusetts, is older than most people who have undergone deep brain stimulation as a treatment for Parkinson's. I visited her one day in her home on Washington Street, where she'd lived for forty years. Her husband, a civil engineer, had died two years earlier, after fifty years of marriage.

Nancy was diagnosed with Parkinson's in 2001. Her hand-writing, she says, gave her away. "I worked in a bank; I was in customer service. My handwriting has always been beautiful—I learned to do it in Catholic school—and a lot of times the bank would ask me to write a note because my handwriting was so nice. All of a sudden the beautiful handwriting was gone." She also noticed that after she stood up at her desk, she would stand still for a few seconds before launching into her stride—a classic example of movement hesitation and a symptom of Parkinson's.

Nancy's doctor put her on Sinemet, the commercial label for-mulation of carbidopa-levodopa. And like thousands before her, she found that it initially controlled her symptoms. But after two to three years, the honeymoon ended. She developed an "allergy" to the drug. She says that she started "bobbing and weaving even when taking a quarter of a tablet." While at times she moved too much, on other occasions the Sinemet would suddenly stop working, and she would freeze up. She's tried to spread out her medication—taking seven small doses of Sinemet throughout the day—but she still experiences motor complications. Over the course of an hour, I saw her go from a rigid state to one in which her upper body oscillated violently.

"I really have no life," she told me. "I can't leave home. I had to give up driving . . . I can change in an instant. I can be walking or I can be in a restaurant, and if I freeze up, it's very difficult to get out of the restaurant or continue my journey." She said that she recently had a number of falls and accidents "caused by the motion of the dyskinesias." After one particularly painful fall, she broke her leg and was referred to a rehabilitation center. Another consequence of the persistent dyskinesias, which sometimes lasted all day, was weight loss. Says Nancy, "I ended up at ninety-five pounds." These events, combined with the urging of her three children, Larry, Laura, and Lynn, persuaded her to seek DBS. Even though she was seventy-six, her doctors saw no sign of cognitive impairment and considered her healthy enough to endure the procedure.

Deciding to undergo deep brain stimulation is a difficult choice for a Parkie to make. Results vary because patients vary in pathology and surgeons vary in skill. It's expensive—usually between $100,000 and $150,000 for the surgery and follow-up—but fortunately Medicare and most private insurance companies, including Nancy's HMO, cover it for "approved" patients. In addition, patients face not insignificant risks of morbidity and mortality, including intracerebral hemorrhage (about 2 percent), infection (about 4 percent), and hardware failures such as breaking leads and electric shorts (about 4.5 percent).

When Nancy talked with me about the risks and the benefits of DBS, rather than tell her about Toby Govea, I highlighted the case of another one of the first Americans to undergo DBS, the journalist Joel Havemann. Havemann was diagnosed with Parkinson's disease in 1990. In 2004, after fourteen years with Parkinson's, Joel—who by then had disabling tremors and dyskinesias—elected to have DBS, still a relatively new operation. DBS not only relieved his symptoms but also enabled him to reduce his intake of L-dopa.

In November 2004, shortly after his operation, Havemann wrote, "I'm no longer a slave to the levodopa that I had to take every two to three hours . . . I can go to a three-hour show without worrying that I'll have to leave before it's over . . . If I can stumble through the rest of my days in about the shape I'm in now, that will be just fine with me." Nearly a decade later, when I spoke to him on the phone, Joel told me that those benefits had lasted: he still has no tremor, dyskinesia, or rigidity, and his L-dopa medication has remained one-third of what it was before DBS. I'd told Nancy he has no doubt he'd done the right thing. In Joel's words, "If I hadn't done DBS, I'd be dead."

Nancy, of course, found the story encouraging. "I'm hoping the dyskinesias will subside so I can get back my driver's license and be able to go down and have my hair done." I was rooting for Nancy, because DBS was the only possibility for her to recover some independence and because I wanted to believe in it as an option down the road for me. So on a snowy March morning in 2013, I arrived at the Beth Israel Deaconess Medical Center West Campus to witness Nancy's operation.

Inside, despite the hour, the corridors swarm with surgeons, residents, and nurses in blue scrubs, moving with urgency. After clearing security and changing into scrubs, I enter the operating room. Half a dozen medical staff members focus on their specific responsibilities, connecting tubes, laying out equipment, reviewing protocols. At the center of the room, Nancy lies awake on a gurney, a little groggy and fighting the urge to sleep. It's essential that she stay alert through today's operation because she will be helping Beth Israel Deaconess's chief of neurosurgery, Ron Alterman, and his team to navigate her brain. Throughout the course of the operation, they'll ask her questions and observe her behavior.

Alterman and his surgical fellows position themselves to work behind Nancy's head in the sterile area. Nurses have loaded the surgical table with instruments, including a pneumatic drill with miscellaneous bits and what looks like a brass sextant. To Nancy's right, the anesthesiologist sits by a bank of machines that flash and beep. On the left side of her gurney stands the team's neurologist and codirector of Beth Israel Deaconess's deep brain stimulation program, Ludy Shih.

Building on the pioneering work of Mahlon DeLong, Alim-

Louis Benabid, and Jean Siegfried, over the next three hours the team will insert a stimulating electrode into the right side of Nancy's brain to mitigate the symptoms on the left side of the body. If all goes well, in three weeks the team will insert a second electrode in the other side of her brain. Younger patients who have the stamina often have both sides done at once (as happened with Toby Govea), a procedure that takes six hours.

Nancy doesn't look her best. Already that morning, Alterman has numbed the surface of her skull, fixed a metal frame to the back of her head, and sent her for an MRI. He's shaving her head and painting it with purple antiseptic liquid. On top of everything, Nancy hasn't had any L-dopa medication since she woke up at 4:00 a.m.—so her drug-induced dyskinesias will not interfere with the operation—and this has left her as rigid as a board.

Ron Alterman, the captain of today's mission, seems confident. He's done more than one thousand DBS procedures, for Parkinson's and also for childhood dystonia, a movement disorder involving prolonged muscle contractions leading to abnormal postures. In his fifties, Alterman recently moved to Boston from New York City. He speaks in punchy phrases, which convey a sharp intellect and limited patience for slow-witted listeners. Alterman, like Dr. Langston, believes location is everything in DBS surgery. The key to success is finding exactly the right place in the brain. Today's target, the subthalamic nucleus—located just above the substantia nigra—is small. "How small?" I ask. "It's about the size of a small olive," says Alterman. "It's located seven to eight centimeters below the surface of the brain. We're securing it with a cannula [a small guide tube] the size of a toothpick."

To pull off such navigational precision involves a blend of digital and analog technology. Alterman strolls over to a monitor on a table by the door and reviews Nancy's digital MRI scan. He chats with his surgical fellow about the best path to Nancy's subthalamic nucleus, one that will be the least likely to "nick the ventricle." The computer does the math and comes up with

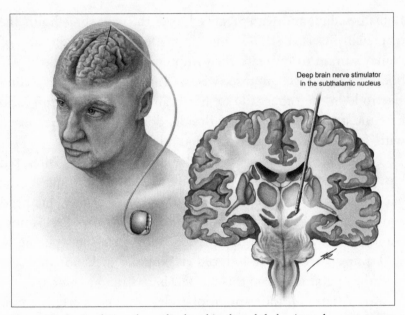

Deep brain nerve stimulator
in the subthalamic nucleus

Figure 10: A stimulating electrode placed in the subthalamic nucleus (Copyright ©
Marie Rossettie, CMI)

a geometric solution connecting an entry point in the back of
Nancy's skull to the subthalamic nucleus. With his course plotted, Alterman exits the OR, scrubs up, and dons his operating
gloves and gown. He returns, walks to the sterile instrument tray,
and assembles pieces of the sextant, known by surgeons as the
Leksell stereotactic frame. Named after the neurosurgeon Lars
Leksell, of Sweden's Karolinska Institute, this rather beautiful
hemispherical instrument is a 3-D reference frame that helps neurosurgeons reach any point in the brain through a small hole in
the skull. Stereotactic surgery has been likened to the task of targeting a submarine hidden deep underwater armed only with
the submarine's depth, latitude, and longitude. If all three are
known with sufficient precision, then, given the proper aiming
device, the submarine can be hit, even though it can't be seen
from above the surface at all. Stereotactic surgery similarly seeks
to target a specific brain region that cannot be seen, by determining

the target region's precise spatial coordinates. Alterman mounts the instrument on the metal frame fixed to Nancy's head. With the device set to the correct coordinates for latitude, longitude, and depth, he is now ready to place the tip of a needle in the subthalamic nucleus.

Alterman drills a dime-sized hole through Nancy's shaved head. The sound is grating. Bone shavings fly like sawdust. He clears away the flakes. Now he has direct access to his patient's brain. Nancy, who is only mildly sedated, is very chatty. But Alterman mostly keeps quiet. At one point, she asks the nurse, "Is he mad at me?" "No," Alterman says, breaking his concentration, "I'm not mad at you."

Now comes the critical part: finding that olive-sized target. The computer has laid out the route, the Leksell reference frame guides their entry into the brain, and the surgical team will verify their position by *listening*. "Nancy, we're just going to listen to your brain. Is that okay?" Alterman asks his patient. Nancy is unfazed. "Yes, that's okay," she replies. The neurologist Ludy Shih smiles and stands close to Nancy to warn her that part of her neurological assessment will involve mobilizing her arms and legs. "I'll be moving you around. Is that okay?" "That's okay," says Nancy.

Alterman orders all the lights turned off. The room's houselights and ceiling-mounted illumination are shut down, leaving only the glow of the bank of machines monitoring Nancy's vital signs and the light from a few computer monitors. Alterman explains why. "To listen to the neurons, we have to amplify their signals about eight thousand times, and the hum of the 60-hertz mains frequency may interfere with our recordings of the neuronal activity."

He inserts a cannula into Nancy's brain, following the computed path. Then he slides a tiny microelectrode inside this guide tube and positions it so it's five millimeters above the target, the subthalamic nucleus. Then the team listens.

We hear a sound that's a bit like ocean waves, and we watch the waveform on a computer screen. This soft sound is white matter nerve cells firing, amplified thousands of times. The atmosphere in the room is tense. As the electrode is lowered millimeter by millimeter toward Nancy's subthalamic nucleus, everyone listens intently for a change in the sound. Finally, and suddenly, the sound morphs. It gets prickly. We hear individual firings, then lots of discharges together like the chatter of a Geiger counter. It's easy to believe these neurons are overexcited. They sound angry. Alterman says they've reached the boundary of the subthalamic nucleus. The team continues slowly moving the electrode deeper into the STN, listening all the time. Eventually, a new firing pattern emerges. This, Alterman tells me, is no longer the subthalamic nucleus; it's an adjacent nucleus that's part of the substantia nigra, and he wants to avoid it.

Having mapped the position of the subthalamic nucleus, Alterman retracts the recording electrode from the cannula and in its place slides in the stimulating electrode. He carefully positions the device so its four contacts are distributed along the length of the subthalamic nucleus. Then he connects it to a pulse generator.

Shih, the neurologist, is holding her patient's arm. She reports that it's stiff and difficult to move. Now comes the moment of truth. She turns on the pulse generator, and magically Nancy's rigidity melts away. Shih reports she can now move Nancy's arm with ease. That's a good sign, she says, that they have the right spot. "Are you okay, Nancy?" Shih asks. "I'm fine," Nancy replies.

Comfortable that the stimulating electrode is in the right place, Alterman and Shih now look for potential adverse effects. "The subthalamic nucleus is surrounded by critical structures," Shih tells me. "So we test for any unwanted effects by increasing the voltage. So we don't want any pulling of the face, we don't want to get any painful tingling feeling, we don't want any voice problems." It's now obvious why they need Nancy awake. They boost the voltage to four volts, and Nancy, who is fully lucid, reports feeling mild tingling. "How about pulling of your face?" asks Shih. "No, my face feels fine," replies Nancy. This "pulling of the face," Alterman says, is the most common warning sign that the electrode placement needs adjustment. "Usually, if I get pulling of the face, I have to move the electrode back one millimeter or so . . . On the other hand, painful tingling suggests I'm a little too posterior, and I need to move it forward a little bit." Shih asks Nancy to say the days of the week, and she does so with no trouble. Her voice seems unaffected. After running all their usual checks, Alterman and Shih announce they are satisfied. The electrode is well placed. The first phase of Nancy's DBS procedure is complete.

Lights come back on. The operation winds down. Alterman closes his patient's small wound. Nancy looks battered—she's bald and purple, after all—but she's passed the first milestone. She has to come back to the hospital in three to four weeks to have the second electrode along with a pulse generator and battery inserted in her chest and then return a month later to have it turned on and programmed. Dr. Shih explains that it usually takes two or three additional sessions to adjust the device so that the benefits are optimal for the patient.

I felt pleased that Nancy was off to a good start. There was no guarantee that her case would turn out as well as Joel's or Toby's. Results of deep brain stimulation to the subthalamic nucleus are variable—ranging from spectacular to complete failure. Given how difficult it is to determine precisely where to place the electrode, however, I didn't find this surprising.

About two months later, I arrived for Nancy's follow-up appointment at Beth Israel. Now that both surgeries were complete, it was time to activate and adjust the pulse generator. Dr. Shih greeted me with a smile. While we wait for Nancy, we speak about the work of the center. Mostly, she tells me, they use DBS for

Parkinson's, but the technology is also used to treat the childhood dystonia that Alterman mentioned, adult balance problems, intractable depression, and even obsessive-compulsive disorder. Surgeons target different brain regions for these diverse disorders and use different voltage and frequency settings, discovered through trial and error.

With Parkinson's, she says, the benefits generally flow immediately after the implantable pulse generator is turned on. With some disorders, like childhood dystonia, the benefits of DBS take months to kick in. Shih had just coauthored a fascinating paper involving a rare side effect of Parkinson's called Pisa syndrome, where a patient leans to one side when walking. The video accompanying the article is astonishing. It shows a sixty-two-year-old woman lumbering down the hall, and her body is inclined at about twenty degrees to the vertical. This is the "before treatment" condition of the patient.

In the study, Shih and her colleagues targeted DBS on a region in the woman's brain located beneath the substantia nigra called the pedunculopontine nucleus. After a week, the woman showed no change. After thirteen weeks, she still leaned to one side. But the team kept the stimulation going for more than a year. At fifty-one weeks, the video shows the woman walking upright, almost normally. And at a little over a year, the problem had been solved. The researchers have theories related to such clinical successes but really no clear understanding of why it worked. DBS has a hit-or-miss quality about it that is both alarming and exhilarating: alarming because experts like Shih and Alterman freely admit their ignorance, and exhilarating because the successes are real and dramatic.

Shih and I walk to the exam room where Nancy is waiting with her son, Larry. She's pleased to see me; I think she regards me as a source of unbiased information. "How are you feeling?" I ask. "Oh, much better," Nancy replies. "I was pretty sore after the second surgery. But I'm good today." It is remarkable to think that this seventy-six-year-old woman has survived two brain surgeries (only one of which I attended). She now has a thin metal "lead" (about one millimeter thick) in each side of her brain, positioned in the subthalamic nucleus region of her basal ganglia. Wires run from both leads under the skin of her face and neck into two cavities in her chest, each containing the implantable pulse generator that Alterman and his colleagues installed. All the hardware is in place. Today, Dr. Shih will turn it on and program it.

There's something odd, I realize. Nancy has taken no medications since 7:00 the night before. Her device hasn't been activated yet. Yet Nancy is not frozen. In fact, she's moving quite well—well enough to go through a full exam involving hand, arm, and leg exercises and standing and walking tasks. "It could be a placebo effect," Shih speculates when I ask her later. "Surgical placebos can last months. Or it could be that the actual lesion created by the electrode releases neurochemicals and produces beneficial effects." Both explanations sound a little wishful to me. If Nancy's dopamine cells are mostly gone, how can some metal hardware enable her to move? I resolve to find out more about this as I continue my research.

Shih announces she is ready to start programming. More precisely, she says she will "map the electrodes," first on one side, then on the other, which means she will test out each contact in turn. She uses a cuplike device to activate (by induction) the stimulator in the right side of Nancy's chest, wired to the right side of her brain; that's the hemisphere that controls the *left* side of her body. Shih starts with contact 0, the one buried deepest in Nancy's brain, the contact that produces the largest effects. Her purpose is to find a voltage level that improves Nancy's ability to move her limbs, but not a level so high as to cause adverse effects like muscle spasms and tingling sensations.

Shih moves the dial, sending a pulsing electromagnetic field through part of Nancy's brain. She boosts the voltage to three and a half volts.

"Whoo!" says Nancy. "I've got vibrations in my leg, and it's spreading to my arm."

Shih lowers the voltage.

"Is it still there?"

"No, it's gone," says Nancy.

"Let's see what happens now," says Shih.

"Now it's only in the hand," says Nancy.

"Where is it now?" says Shih.

"Oh, now it's in my face," Nancy announces.

"I'll turn it down," Shih says.

When she lowers the voltage to two volts, the side effects disappear. So now Shih tests Nancy's muscle tone—there's no rigidity—and measures how well she moves. She is constantly stopping to take notes.

Dr. Shih methodically moves on to the other contacts (1, 2, and 3). Again, she is looking for the sweet spot, a voltage setting where Nancy's symptoms improve significantly but where there are no adverse effects—no tingling skin, face pulling, eyes closing, feelings of fatigue, voice problems, and more. This is similar to what Shih and Alterman described to me when they tested the electrode in Nancy's DBS operation. Contacts 2 and 3 seem to have fewer adverse effects on Nancy's body, although at one point she says, "I just feel kind of fuzzy." But the stimulation going on deep in her brain is clearly working. Nancy can lift her left leg much higher than before, and she is able to open and shut her hands very rapidly (another test that indicates the severity of her motor symptoms).

Shih repeats the mapping process for the four contacts on the left side of Nancy's brain (the hemisphere controlling the *right* side of her patient's body). Again Nancy experiences some troublesome side effects, at one point reporting that her eyes are partially shut, and Shih lowers the voltage. After an hour of adjustments, the civil war in Nancy's brain seems much less intense. Nancy's wiggly right leg no longer wiggles. The dyskinesias are gone.

Shih announces the final parameters for this first programming session. She selects contact 2, with a voltage of two volts, on both sides. She hasn't touched the frequency (130 hertz) or the pulse duration (sixty-millionths of a second). According to Shih, those parameters don't make very much difference in Parkinson's. But other neurologists I have spoken with argue that adjusting the vibrational frequency can fine-tune symptoms: raising it slightly can help with tremor; lowering it slightly can help with freezing of gait. The stimulator can be used in monopolar mode, where the battery is the positive terminal, or dipolar mode, where two adjacent contacts are the electrical poles. The dipolar mode, some neurologists argue, allows a tighter focus, which enables some adverse effects to be eliminated. Some advanced stimulators are switchable between configurations that spare or enhance particular functions (like the voice) when needed.

After I say goodbye to Nancy, Larry, and Dr. Shih, I reflect on what I observed. While there was definitely something going on, neurosurgeons and neurologists had so little understanding of the underlying mechanisms that it felt a bit like black magic to me.

Neuroscientists are working on next-generation technologies that might bring enhanced understanding of the mechanisms and also offer greater benefits. Instead of implanting devices that simply deliver a continuous electrical stimulation, they are developing technologies that deliver stimulating jolts only when required. Modern cardiac pacemakers can do this, so why not DBS "brain pacemakers"? The idea is to design DBS so that the system can monitor the electrical activity in the basal ganglia, and when it detects an abnormal signal, it can respond automatically with an appropriate stimulation. A smart device—a so-called neural network modulator—could in theory respond to the patient's changing brain and deliver a more personalized therapy. In April

2015, the University of California, San Francisco, neurosurgeon Phil Starr reported that his team had demonstrated this concept in a cohort of twenty-three DBS patients. Hypothesizing that Parkinson's patients' motor problems result in part from excessive synchronization between brain circuits in the basal ganglia and the motor cortex, the UCSF team showed that they could detect episodes of lockstep cortical synchronization and activate the DBS device on demand, thereby diminishing the unwanted electrical behavior.

When I called Nancy to check on her condition a couple of days later, she was in good spirits. She had cut her medication by half (taking a quarter tablet of Sinemet four times a day), and the dyskinesias were pretty much gone. She was already putting on weight and sleeping well at night, and she was experiencing no "off" periods during the day.

Over the next year, Nancy returned to Beth Israel several times so that Dr. Shih could tweak her programming. This kind of follow-up to tweak the settings is quite common for people who have undergone DBS. The good news is that Nancy, like many recipients of DBS, continued on the lower dose of medication and the violent dyskinesias never came back (although she still has a bit of wobble in her right leg and right arm). But Nancy discovered, as many patients have, that deep brain stimulation can indeed have costs as well as benefits. When I spoke to her close to a year after her first surgery, her mood was subdued. Nancy told me that she'd developed serious problems with drooling. More significant, I noticed that her voice had changed. When she spoke, she slurred her speech and was difficult to understand. I knew from speaking with clinicians like Alterman and Shih that voice problems are one of the most cited complications of DBS. The subthalamic nucleus sits very close to the corticobulbar tract (the motor pathway controlling speech), and it's not always possible to steer the pulses so that they do good in one area and avoid all harm in another structure that's nearby.

Nancy's life hasn't been transformed in the way she (and I) had hoped. She still spends most of her time at home and will only go out if accompanied by family or friends. And while she tries to keep active, she is worried about living alone in her home for the rest of her life and is considering moving to an assisted-living facility. But Nancy remains upbeat. When I ask her if it was all worth it and if she would recommend deep brain stimulation to someone else, she doesn't hesitate: "Yes, without a doubt. I watched Michael J. Fox on television the other night, he was bouncing and squirming [with dyskinesias], and I thought, 'Oh my God, that was me.' No, I have no regrets. It was worth everything I went through. Oh my God, yes. And my kids agree."

7

THE EXERCISE Rx

You may not be able to slow the progression of the disease,
but you can certainly slow the progression of the disability.
—Terry Ellis, College of Health and Rehabilitation
Sciences, Boston University

My time with Nancy had given me insights into the pros and cons of an invasive therapy—deep brain stimulation. But apart from drugs and surgery, there is another intervention that many researchers are beginning to believe helps tame the symptoms of Parkinson's, and that's exercise. These days, everybody seems to extol the virtues of exercise, promoting activities from spinning to tai chi to kickboxing to yoga, and neurologists are no exception when it comes to Parkinson's. They emphasize the importance of stretching, strength training, and aerobic exercise in particular. There's no danger of side effects like those that patients get from drugs and deep brain stimulation. But so far, there's relatively little hard evidence that such calisthenics can fundamentally change the trajectory of the disease. Professor Terry Ellis is trying to fix that.

Ellis first realized the potential of exercise for Parkinson's patients when she worked as a physical therapist at Boston's

Spaulding Rehabilitation Hospital. Together with the neurologist Robert Feldman—who had observed anecdotally that his Parkinson's patients who exercised fared better than those who didn't—she started developing ideas for investigating exercise therapy scientifically. Ellis went on to do her PhD and is now a leading exercise researcher.

I met Ellis in her fifth-floor office in Boston University's College of Health and Rehabilitation Sciences, from which she directs the Center for Neurorehabilitation. Ellis radiates energy and passion. And the subject she's most excited about is exercise.

Back in 1980, when Ellis worked as a physical therapist, exercise science was in its infancy. There were no controlled trials to assess the benefits of exercise for Parkinson's as there were for the testing of new drugs and devices. Today there are some fifty published Parkinson's exercise studies—which test the effects of everything from walking to tai chi to tango dancing to boxing. They're all short term—most of a few weeks' duration—but they reveal definite benefits from exercise: improvements in physical function, quality of life, cardiovascular fitness, and cognition. Researchers have shown that twenty-four weeks of a twice-weekly tai chi workout, for example, improves Parkinson's patients' stride length and reduces the number of falls. Another research team demonstrated that following twelve months of tango dancing, Parkinson's patients showed improved balance and gait and developed an enhanced ability to multitask. Studies of boxing, an activity that involves rapid movements in different planes, report that this exercise activity improves balance in people with Parkinson's. And epidemiological research also shows that people who engage in moderate to vigorous exercise are less likely to contract Parkinson's disease in the first place.

That's the good news. The bad news is that very few Parkies exercise. In fact, people with Parkinson's tend to become even more sedentary than the general population. Ellis told me about a study in the Netherlands where researchers followed a group of seven hundred Parkinson's patients and compared them with a control group of adults of the same age without Parkinson's. The Dutch researchers found that people with Parkinson's disease were about one-third less active than the control group. But, Ellis adds, "it's worse than that because people without Parkinson's disease are not active either." It's true. Americans, in particular, are couch potatoes. Recent studies show that more than 97 percent of Americans over sixty years old fail to exercise even the minimum recommended amount—150 minutes a week of moderate activity (for example, walking with an average of one hundred steps a minute).

Parkinson's patients face two problems. First, they are battling a progressive disease—one that will, in the long term, likely even disable an Olympic athlete, as it did the New Zealand runner John Walker. In other words, the symptoms can make exercise more and more difficult to perform. Second, this process of physical decline can accelerate if Parkies simply give in and allow their bodies to atrophy.

Ellis has just completed her own study that followed Parkinson's patients over a year, monitoring their activity with special ankle bracelets that counted the number of steps the patients took every day. The results showed that Parkinson's patients did indeed become less active over time. "Over the course of a year, the number of steps taken per day dropped by 12 percent on average," Ellis explained to me. The number of high-intensity steps (done at a rate of one hundred steps a minute or more) dropped by even more—by 40 percent. The implication: Parkies who do nothing *will* go downhill fast; they'll grow old quicker than their unfit non-parkinsonian contemporaries.

Some scientists argue that there's a good reason for this: Parkies, they claim, suffer from a motivational deficit sometimes referred to as abulia—meaning lack of will. The breakdown of the reward system, which is designed to give immediate rewards (bursts of dopamine) for performing successful tasks, can lead to a certain kind of listlessness. But what if Parkies can overcome such apathy and fight back? How much difference can exercise make to their future? It turns out that it could be quite a lot. One of the most compelling and hopeful research stories I have heard to date involves another cycling story.

The star of the tale is Dr. Jay L. Alberts, the Parkinson's researcher and accomplished cyclist. In 2003, he decided to enter the seven-day RAGBRAI bicycle ride across Iowa as part of an effort to bring Parkinson's awareness to the people living in rural areas of the state. Alberts, a native Iowan, planned to assemble a group to ride tandem bikes in the annual event. He asked several friends to join him, including Cathy Frazier, a forty-eight-year-old woman with Parkinson's, and her husband, Ralph. The trip started badly for the couple. Shortly into the ride, Cathy and Ralph—who had little experience on a tandem—started to argue and decided it wasn't for them. So Alberts switched with Ralph and rode with Cathy for the rest of the week. Alberts sat up front—in bike parlance, he was "the captain"—and Cathy, who rode behind, was "the stoker."

According to Alberts, over the next few days, two remarkable events occurred. First, Cathy told him, "I don't feel that I have Parkinson's when I'm on the bike." Second, Alberts recalls that two days into the ride, during a rest break, he happened to see a birthday card that Cathy wrote. The handwriting was beautiful. Prior to the ride, Cathy's handwriting displayed classic micrographia, that cardinal sign of Parkinson's disease whereby patients' handwriting gets smaller and smaller. Now, after two days on the tandem, she could write normally. While the improvements were temporary, Alberts thought he had witnessed something important.

Alberts continued his busy life as a professor at Emory University, but each year when he did the RAGBRAI bike ride with parkinsonian stokers, he noticed similar phenomena. For example, after completing a tandem ride, one tremor-plagued patient said with astonishment, "I can drink with my left hand." Whatever was going on, Alberts conjectured, had to be connected to the fact that the stoker at the back was being pulled along much faster than when she rode solo. Alberts, who has the physique of a quarterback, pedals on average at eighty to ninety revolutions per minute. The stoker's normal cadence is about sixty revolutions per minute—a difference of about 40 percent. In other words, the Parkinson's patient on the back of the bike is under pressure to up his game.

Then, in 2007, it happened that one of the people riding with Alberts had undergone a deep brain stimulation operation on his STN. This rider was about to hop on the tandem when he said to Alberts, "Why don't we really test out your hypothesis—I'll turn my stimulator off." They got on the bike and set out for the fifty-mile ride. "And I'll never forget, we got about fifteen miles into the ride, and we stopped to have a Danish. The DBS patient held the Danish in his hand and said, 'What happened to the tremor in my right hand?' We went on to finish the ride, and at the end, after four hours, his symptoms had largely vanished," Alberts recalls.

Now Alberts was convinced. He started reading the scientific literature and discovered that there was a lot of research on animals that reported similar results. A long tradition of rodent and monkey research dating back some fifteen years had linked intense exercise on exercise wheels and treadmills with improved brain health, including motor abilities and cognitive function, and even indicated a neuroprotective effect against MPTP parkinsonism. Typically, in these experiments, one group of animals is placed on a motorized treadmill that forces the creature to run faster than it wants. A sedentary group of caged animals (the "cage potatoes") serves as a control. Then, after a training interval of a few weeks, both groups are given a neurotoxin such as MPTP. The sedentary group suffers immediate damage to the substantia nigra and develops parkinsonism; the exercise group partially resists the toxin.

Paradoxically, research on humans' exercising had not found comparable benefits. And Alberts thought he knew why. "The human experiments generally test voluntary exercise," Alberts told me, "but the rats are doing 'forced exercise.' They are forced to run at a faster rate—and given electric shocks if they don't." And that's basically what was going on with the tandem bikes. When patients like Cathy were riding on their own, they pedaled at fifty to sixty revolutions per minute, whereas when they were paired with Jay on the tandem bike, they moved along at eighty to ninety revolutions per minute. "So, in essence, I was forcing them to pedal at a higher rate."

How did he know the parkinsonian stokers weren't just coasting? Alberts insists, "I could feel them working and feel them contributing." To prove this, he and his colleagues built a special tandem bike that could measure the relative power contributed by captain and stoker. It turned out that the rider in the front was contributing 70 to 75 percent of the power, whereas the Parkinson's patient at the back was inputting some 25 to 30 percent of the power.

Just as with the rats on a treadmill, forced cycling, Alberts hypothesized, improves brain function and with it the symptoms of Parkinson's disease. It's probably, he says, a combination of "the quantity of information [that's increased as the rider pedals faster] and the consistency of information that is sent to the brain."

An intriguing theory, but did it hold water? To test it, Alberts—who by now had moved to the Cleveland Clinic, where he was director of its Concussion Center and a staff member in its Department of Biomedical Engineering—carried out a pilot trial that took ten Parkinson's patients (on average seven years into the disease) and randomly assigned them to one of two groups, each with five people: a treatment group receiving forced exercise and a control group voluntarily exercising. The forced-exercise group was given three sixty-minute workouts on a tandem every week for eight weeks, in which a trainer-captain pedaled at eight to ninety revolutions per minute. The control group did three sixty-minute sessions each week on a bike by themselves. Each group was evaluated before and after the experiment—by clinicians blind to which group the patient was in—using the Unified Parkinson's Disease Rating Scale (UPDRS) as an outcome measure.

Every person with Parkinson's knows this test, which is the centerpiece of a neurological exam. The widely used motor component of the UPDRS, which takes about fifteen minutes, covers fourteen categories where muscles may be behaving badly, including speech, facial expression, tremor, rigidity, finger tapping, hand movements, hand pronation/supination, foot tapping, ability to rise from a chair, posture, gait, and balance. The aggregate score (between 0 and 108) is supposed to quantify a patient's motor condition and, over time, chart the progression of the disease.

The scale has many critics. While some neurologists and researchers defend it, patients privately mock it as an unrepresentative snapshot of their physical states. As the Swedish patient and advocate Sara Riggare puts it, "I see my neurologist every six months for a thirty-minute session and thus spend every year a total of one hour being observed [with the UPDRS]." That leaves, she says, "8,765 hours per year when my disease is not being monitored." The test is highly subjective as well. Two neurologists will often give a different score for the same patient on the same subtest. Results vary dramatically depending on how recently patients have taken L-dopa and the time of day. Patients perform up to 30 percent better in the clinic than they do at home. Yet, with all its flaws, everyone—from researchers like Jay Alberts to heads of pharmaceutical companies—is compelled to use it as the controlling metric.

Given the UPDRS's limitations, what did Alberts notice?

After eight weeks, the forced-exercise group's symptoms had dramatically improved—reducing their average UPDRS part 3, or UPDRS3, score (measured when they were off dopaminergic medication) by about one-third. That's almost as large an effect as the one brought about by levodopa. By contrast, the control group saw no benefit. Two weeks later, these gains were still there in the tandem group. But four weeks after treatment, the forced-exercise group was back where they'd started. In other words, you have to keep exercising. This, Alberts says, is exactly the same for drugs. "When would you ever prescribe medicine for eight weeks and then say you're done, you don't need to take it again? Obviously, you have to keep doing the exercise, and that's a good thing."

Because it's impractical for every patient to have permanent access to a trainer-captain, Alberts is trying to reproduce the experience on a motorized bike. Then he will have to prove in a larger, longer-term study that it delivers benefits in a reliable manner.

His concept of forced exercise is just one of many options out there that claim to help Parkinson's patients. What's the hard evidence that it works better than, say, tai chi, kickboxing, progressive strength exercises, or high-impact cardio in slowing the disability? So far, not much. Exercise research is not only methodologically tricky but also expensive to do—especially for long-term studies. Currently, exercise studies (like most drug studies) last from a few weeks to a few months. It's very difficult to detect an effect in such a short time, especially given the crudeness and subjectivity of the clinical assessment scale (the UPDRS), the confounding influence of levodopa, and the variability in patient symptoms from day to day.

But what specific exercise you choose may be less important than selecting an activity that will address your specific needs and be sustainable. Traditionally, the clinicians most interested in developing therapeutic exercise programs have been physical therapists, who dominate rehabilitation services for movement and gait disorders. The physical therapist Dr. Fay Horak, who runs a balance laboratory at the Oregon Health and Science University in Portland, says that neurologists were originally skeptical about rehabilitation. "When I first started, they wouldn't even talk to anyone who wasn't a neurologist. They said, 'Parkinson's disease is a degenerative disease—why would I send them to rehabilitation?'"

One exception to this trend—a neurologist who has long been supportive of the importance of physical therapy—is the Dutch researcher Bastiaan Bloem, the neurologist whose patient could cycle but not walk. Before training as a clinician, Bloem considered a career as a professional volleyball player. He was good enough, and he was certainly tall enough. But in the end, for personal reasons, he opted for medicine. "I chose to go to medical school," says Bloem, "because my mother had multiple sclerosis. From my first days as a child, I saw firsthand what neurological disease was like. My dream was to cure my mother."

Bloem, who resembles a younger, taller, and more disheveled version of the actor Sam Shepard, did go on to make a major contribution to the treatment of neurodegenerative disease, but it turned out to be in Parkinson's rather than in multiple sclerosis.

As Bloem finished his neurological training, he became convinced of one central truth. While the classic parkinsonian symptoms of rigidity, tremor, and bradykinesia are mediated by dopaminergic systems in the brain and respond to levodopa treatment, this wasn't true for posture and balance deficits. Those disabling symptoms involve damage to other brain regions mediated by different neurotransmitters and, logically, might be amenable to other therapeutic interventions.

As a volleyball player, Bloem had developed a huge interest and belief in physical therapy (which is often referred to as physiotherapy in Europe). Says Bloem, "I was eager to demonstrate that physiotherapy—which is based on the idea of helping patients to compensate for their disability—would be a better treatment of gait and balance than just L-dopa." What happened next was pure coincidence. Bloem joined a statistics course being held on one of the small islands in the north of Holland. As it happened, his roommate was a physical therapist and scientist named Marten Munneke. As Bloem recalls, "Students were given the task of designing a trial, so Marten and I said, 'Why don't we design one for Parkinson's disease and the influence of physiotherapy?'"

Almost immediately, they saw a problem. Very few Dutch physical therapists were trained to work with Parkinson's patients. "Back then, physiotherapy in Holland basically reminded us of homeopathy: care was strongly diluted, with every therapist typically treating only a few Parkinson's patients, and this kept them from acquiring sufficient expertise with this complex disease. Parkinson's patients complained that the emphasis was on drugs and deep brain stimulation, and rehabilitation was neglected." So Bloem and Munneke decided that before they could run their trial, they would have to set up a program to train physical therapists and other health professionals.

This led to the radical idea of ParkinsonNet. Instead of neurologists' delivering episodic care in hospital settings, Bloem and Munneke (who had become firm friends) set out to build a national network of highly trained health professionals capable of treating Parkinson's patients in the hospital, the community, and the home. ParkinsonNet started in 2004 in the cities of Nijmegen and Arnhem with 19 physical therapists, 9 occupational therapists, and 9 speech therapists. Today, with sixty-six regional networks, 2,970 trained professionals (including neurologists, physical therapists, pharmacists, psychiatrists, and occupational therapists), and some fifty thousand patients, the program covers the whole of the Netherlands.

To start with, it was a hard sell. Bloem recalls, "At the beginning, many neurologists were reluctant to participate. They didn't believe in the merits of allied health interventions, and in truth they had a point, because there actually wasn't much evidence for it." But after a decade of use, there are hard data. In the March 19, 2014, online edition of the *British Medical Journal*, Bloem and Munneke summarize what ParkinsonNet has accomplished. By treating patients using trained experts in the community rather than hospitals, ParkinsonNet saves some $28 million annually. It's also produced better health for patients—one large observational study found a 55 percent decrease in hip fractures in cohorts treated in ParkinsonNet regions. For Bloem, it's an example of a new collaborative culture of care where "specialized professionals and engaged patients work together to try to achieve optimal outcomes."

This collaborative culture has not been an easy sell in the United States. Dr. Fay Horak, who like Professor Terry Ellis started out as a humble physical therapist, is today one of the world's leading experts on balance. She runs a human performance laboratory at the Oregon Health and Science University in Portland that researches the various balance challenges Parkinson's patients face: from the flexed posture to the reduced ability to recover from being pulled off-balance. Because poor balance leads to falls—a major source of morbidity among Parkinson's patients—Horak wants to do everything to keep them safely on their feet, including offering an "agility boot camp."

Horak says that the human gait is exquisitely revealing. "You can see Parkinson's disease in a person's gait even before you can detect it with a standard clinical exam such as the UPDRS. We can see the upper body doesn't rotate as much, we can see the arms don't swing as much, we can measure a drop in the stride length, and the stride velocity, and we can do this even before people are diagnosed with Parkinson's."

As Bloem and Munneke had shown in Holland, physical
therapy offers a way for patients to compensate for these gait
and balance challenges. For her part, Horak is very interested in
turning. People, she says, turn about a thousand times a day or
more, and she can measure how many steps they take when they
turn. "Turning is harder than walking . . . People with Parkinson's fall a lot more often when they are turning than when they
are just walking." That's because turning involves shifting your
weight by just the right amount at the same time as you perform
an asymmetrical leg motion. Says Horak, "When you want to
turn left, you must shift your weight to the right. It's really
tricky . . . People with Parkinson's disease end up with too much
weight on the wrong leg, and then they fall."

But physical therapists, like neurologists, only occasionally
observe Parkinson's patients. Most of the time, the ups and downs
of the disease are only witnessed by the patients themselves and
their caregivers. About a decade and a half ago, researchers began wondering if technology could help in capturing a more complete picture of Parkinson's, one that tracked the variable symptoms
of patients' going about their daily lives. One of these individuals was the former Intel CEO Andy Grove, who developed Parkinson's in 1999. In an article for *Forbes*, Grove bluntly dismissed
the UPDRS as "a piece of crap." Grove used his money, technical
knowledge, and connections to develop a machine (the At Home
Box) to objectively measure a patient's state. Objective monitoring turned out to be a hard problem to solve, even for a micro-
chip engineer. But at the start of the twenty-first century, Rush
University's Christopher Goetz tested an early version of the At
Home Box and found that it could detect a decline in the UPDRS
and see it before a patient's regular office visit.

Since then, objective Parkinson's disease monitoring technology has become smaller, faster, and cheaper. Over the past
decade, a number of companies have been quietly developing

technologies capable of tracking Parkinson's patients' symptoms outside the clinic 24/7—including the Cleveland-based Great Lakes NeuroTechnologies, the Australian company Global Kinetics, Portland-based APDM, and the Cure Parkinson's Trust–supported European SENSE-PARK project. These entities are working on various combinations of advanced wearable sensors (worn on sites like the wrist, waist, and ankle), which use accelerometers to track multiple domains round the clock—bradykinesia, tremor, walking, gait, balance, cognition, and more. The Great Lakes NeuroTechnologies researcher Dustin Heldman says the devices his company has built are already more sensitive than anything we could find out from the UPDRS. Says Heldman, "In the clinic, it often takes months to years to detect a significant change, but with this device we can detect changes much sooner by monitoring people at home."

To learn more, I tested two systems. One technology, Kinesia HomeView (designed by Great Lakes NeuroTechnologies), consists of a small computer and a finger sensor. Sitting in front of the screen, I performed a set of exercises to order—a kind of modified UPDRS. It took about five minutes. I then pushed a button, and the laptop transmitted the data wirelessly to a server, where the data were analyzed and formatted into a simple table listing thirteen metrics. These numbers estimated the severity of three forms of tremor, gave a measure of dyskinesia, and determined the speed, amplitude, and rhythm displayed during the finger-tapping, hand-movement, and hand-flipping exercises. The report represented a snapshot of my motor symptoms at that particular moment in a form that was easy and convenient for me and my neurologist to digest. The main advantage is that it can be used at any time of day. This device can, therefore, reveal any changes in motor performance over time, providing a much more detailed picture of my disease than that obtained in an evaluation every six months.

The other system I tested was a set of sensors being developed by Portland-based APDM. For this experiment, I wore three small devices, one on my wrist and one on each of my ankles. Each sensor incorporated miniaturized accelerometers, gyroscopes, and magnetometers (a magnetometer is an electronic compass that determines which direction you're moving). The wrist sensor detects tremor and dyskinesia. The ankle sensors are designed to measure multiple aspects of walking. It turns out that walking is enormously complex. There are two basic parts to this everyday act: the stance phase, when at least one of the feet is in contact with the ground (this happens about 60 percent of the time), and the swing phase, when that foot is in the air (about 40 percent of the time). The stance phase on, say, the left leg starts with the left heel landing on the ground. The foot then flattens, and then the heel rises until the left toe lifts off into the air. This event, called "toe off," is the start of the swing phase. Coordination of the two legs is obviously crucial to successful walking. When the left leg executes its swing phase, the right leg is in its stance phase (and there is a brief overlap when both feet are on the ground at the same time). As the two legs work together, we move forward with a particular stride length and cadence (the number of steps per minute).

I wore the sensors for three weeks and on most days went for a long walk lasting at least sixty minutes. This activity generated masses of data. After the APDM chief engineer, James McNames, had crunched the numbers, he sent me a report involving some twenty gait metrics. When the data were averaged over time, they revealed emerging asymmetries between my weaker left leg and my less affected right leg. As the graph shows, on average I lifted my left foot higher than my right foot. This makes sense, as it indicates that I was spending more time in each step cycle on my stronger right leg than on my weaker left one. The data also showed that my left foot was not flexing at a sufficiently large angle when it lifted off (at toe off) or when it landed on the heel. Such subtle gait asymmetries, McNames told me, are the first signs of trouble. If not addressed, they might lead to worse gait problems in which I would end up shuffling my feet when I walk rather than landing cleanly on my heels.

I hope that APDM and other companies will go on to develop wearable sensors that have real-time feedback mechanisms—with, for example, an audio alert—that can guide me to walk symmetrically and to place my feet correctly. Such a technology has the potential to prevent people with Parkinson's from developing bad habits that put them at risk for falls and freezing episodes. In the years ahead, I plan to continue experimenting with wearable sensors. With big players like Apple, Samsung, Nike, and Microsoft involved, objective monitoring technology will be an exciting area to watch.

Overall, I find the evidence in favor of exercise convincing, and I plan to use it therapeutically for the rest of my life. My daily routine involves stretching and a fast four-mile walk. The daily stretching routine extends the mobility of my neck, shoulders, hips, and feet. My one-hour power walk with my wife, Yanira, is a version of forced exercise. Trying to keep up with a very fit person is a lot like being on the back of a tandem bicycle. Both individuals try to sync up their motions, and the "captain" constantly adjusts the pace—to boost or reduce the intensity. When I walk alone, I use a metronome or music track to drive the pace, something I learned from the dancer Pamela Quinn.

Every few months, I drop by Terry Ellis's center to do the six-minute walk test—a test of whether a walker can maintain a pace of more than a hundred steps per minute. It turns out that gait speed is highly correlated with risk of mortality and that individuals with a preferred walking speed of greater than 1 meter per second (2.24 miles per hour) live longer on average than those with lower speeds. While so far I can comfortably maintain a fast

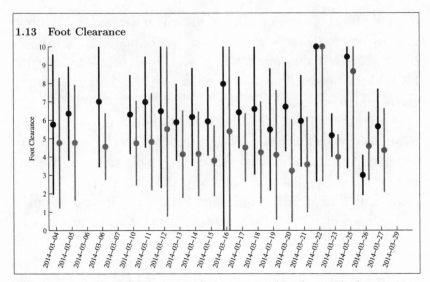

Figure 11: A graph averaging sensor data over several weeks reveals that my more affected left foot (shown in black) lifts higher than my right foot (shown in gray). (APDM, Inc. [Mobility Lab system])

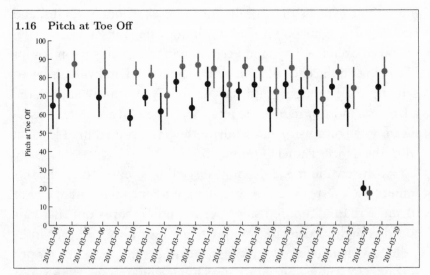

Figure 12: A graph averaging sensor data over several weeks reveals that the angle my foot makes at the moment of toe off is smaller on my left side (shown in black) than on my right (shown in gray). (APDM, Inc. [Mobility Lab system])

walking speed, the conventional wisdom says that in time my disease will overwhelm my conscious will. That's the limitation of symptomatic treatments. What every Parkie hopes for are interventions that modify the disease itself. It is to this quest that we now turn.

8

NEW NEURONS FOR OLD

The Swedish neuroscientist Patrik Brundin keeps a picture of his father on his office wall as a reminder of why he became a Parkinson's researcher. In 1974, when the Brundin family lived in Darlington, in the north of England, a local neurologist by the name of Dr. Saunders diagnosed Patrik's father, Bertil, with Parkinson's disease. Bertil Brundin—an executive for a Swedish lawn-mower company called Flymo—was put on L-dopa, which had just been released in the United Kingdom. Patrik, then twelve years old, remembers his mother asking the neurologist, "Will it arrest the disease?" To which Dr. Saunders blushed and said, "I'm sorry, but we don't know anything about this yet."

The young Patrik Brundin resolved that he would devote his life to finding a cure for his father's disease and in 1980 elected to study medicine at Sweden's celebrated Lund University. There, by luck, he met Anders Björklund, a pioneer of neural transplantation, who also taught histology to the first-year medical students. Within months, Björklund had become Brundin's mentor, and from that moment, Brundin recalls, "I never looked back." He took a break to do a PhD in the middle of his medical studies, joining Björklund to work on a remarkable series of neural tissue grafts aimed at reversing the symptoms of Parkinson's disease. The operations were highly controversial around the world because the transplanted neural tissue came from aborted fetuses of between six and eight weeks' gestation—so controversial, in fact, that in 1988 President Ronald Reagan instituted a federal funding moratorium for any such research based in the United States. (This moratorium was eventually repealed by President Bill Clinton on January 23, 1993.)

The first of the Swedish operations took place in Lund Hospital in November 1987. The surgeon Stig Rehncrona prepared patient 1, a forty-seven-year-old woman, for surgery. Patient 1 had contracted Parkinson's at age thirty-three and could now move for only a few hours each day. Patrik Brundin sat in another room peering through a microscope and dissecting the fetal tissue to be used in the transplant. Research had shown that the optimal time to transplant human fetal-brain tissue from the substantia nigra was from six to eight weeks following fertilization. Go beyond that period, to ten weeks after fertilization, and very few cells survive. It was a delicate process. The entire fetus at eight weeks is only the size of a fingernail; the substantia nigra, the size of the head of a pin. Brundin's task was to dissect only those cells whose destiny was to make dopamine. Very closely attached to the substantia nigra tissue were cells that had a different fate—to become cartilage, bone, skin. Brundin knew from animal experiments that if he wasn't careful and grafted this tissue into the brain along with the substantia nigra cells, they'd grow into big bits of skin and cartilage. So it was critical that these fragments of so-called mesenchymal tissue were completely removed.

After two hours of painstaking dissection, the research team mixed the fetal material from four fetuses with a chemical called trypsin, dissociating the cells into a liquid suspension. Rehncrona then drew this concoction into a cannula and implanted the suspension not into patient 1's substantia nigra but into her striatum. Rehncrona didn't place the cells in the substantia nigra region for two reasons. First, it is a very difficult and dangerous site to reach. Second, there was good reason to believe that grafting to the substantia nigra would be ineffective. Dozens of experiments with rats had shown that cells grafted to this area did not reverse their Parkinson's-like symptoms, because the grafted nerve fibers could not grow such distances to reach their targets in the striatum. In contrast, grafts placed in the striatum itself worked spectacularly well in rats and reversed Parkinson's-like behavior.

Apart from finding the optimal insertion site, the Lund scientists needed to ensure that the grafts contained enough fetal material to make a clinical difference. The team had estimated that only about 10 percent of the fetal cells were likely to survive the process of abortion, dissection, preparation, and insertion and go on to function in the patient's brain. Because of this loss, the team had decided that they would need to implant cells from multiple fetuses. Otherwise there might be no effect at all.

The initial results of the first two cases were modest. But after modifying the size of the cannula, the team began to see positive results. For example, patient 4 was grafted only on one side, and in 1999 a PET scan* of that side showed that the brain region was producing normal levels of dopamine. Remarkably, the patient was off all medication and his condition reversed "by

---

*Positron-emission tomography (PET) is used to detect the accumulation of L-dopa as it's taken up by an area in the brain. A Parkinson's patient is injected with radioactively labeled L-dopa that emits positrons (subatomic antimatter particles, with the same mass as an electron but an opposite charge). The life of a positron is short and explosive. As soon as it meets an electron, its "matter" counterpart with an equal and opposite charge, the two annihilate each other, producing two energetic gamma rays that fly off in opposite directions. In a PET scan, the detectors that encircle the patient are configured only to react to such pairs of gamma rays, emitted simultaneously and traveling in opposite directions. In this manner, the number of gamma-ray pairs is counted and the locations of the associated positron-electron annihilation mapped. The detectors register thousands of events, and a computer integrates them into an image that shows a picture of how much fluorodopa is being taken up in the different regions of the patient's brain—in particular by the striatum, where the axons from the substantia nigra terminate. Before PET scanning, the only way to assess brain dopamine was to acquire a piece of the actual tissue after death.

about ten years," so he could return to an independent life after the surgery.

Over the next decade, a total of eighteen cases using fetal tissue were conducted at Lund. More than three hundred were conducted worldwide, especially in the United States. While some patients experienced little benefit, striking successes, like patient 4, made some researchers suspect that the procedure could succeed when all the conditions were right.

Further evidence that the neural grafts had survived and started pumping out dopamine emerged when two patients grafted in the United States died of causes unrelated to the surgery. Their brain tissue was sent to the Rush University neuroscientist Jeff Kordower in Chicago. Kordower confirmed there was no doubt the transplants had worked. "The grafts," he said, "were huge, tremendous innervation, synapse formation, metabolic enhancement, and so from an anatomical point of view everything we hoped to find we found."

The neural grafting operations continued with encouraging outcomes both in Sweden and in the United States.* By 1999, many people were convinced that this method offered one of the only ways to repair brains heavily depleted of dopamine neurons. But for others, the jury was still out. After all, these were uncontrolled trials—so-called open-label studies. An open-label study is an investigation where both subjects and researchers know which treatment is being delivered. Jeff Kordower recalls, "We had positive patients in open-label trials where the surgery appeared to reduce their need for levodopa dramatically." On the other hand, he continued, "you have to understand what 'open-label' means: there is the potential for a placebo effect."

The placebo effect is a puzzling concept to get your head around. The idea that an essentially inert substance—like a sugar

---

*Among the patients were three of the frozen addicts—George Carillo, Juanita Lopez, and Connie Sainz. I met Patrik Brundin during the making of my documentary for *Nova*, "Brain Transplant," which documented the first two of these transplants into the MPTP patients.

pill or a saline injection—can produce therapeutic benefit is counterintuitive. Yet it captures the important truth that patients can get better for a host of reasons unconnected with a specific intervention. An illness can vary from day to day, and many clear up by themselves. Patients who enter trials are sometimes "different" from those who don't. They might be sicker than other patients. Once they are participating in the trial, they likely get extra clinical sessions with physicians and research staff, in whose company they may behave differently than they usually do. Trial patients may have positive expectations that the new procedure will slow, stop, or reverse their illness. Researchers testing new therapies go to extraordinary lengths to control for such factors—which they refer to as "confounders." They randomly assign patients to either a treatment or a placebo group, keeping both physicians and patients blind as to which subjects actually receive the new agent being tested and which get a placebo version. In this design, the assumption is that the various confounders *should on average affect the treatment and placebo group equally*, and so any difference in clinical outcome can safely be attributed to the novel drug or surgery. If, on the other hand, there is no difference in outcome between the two groups, then the trial is deemed a failure.

While the FDA mandates randomized placebo-controlled trials for new drugs and medical devices, the agency has no jurisdiction over the practice of surgery. And so surgeons seeking to carry out new operations do not have to seek FDA approval. Randomized placebo-controlled trials of surgical procedures are also intrinsically more complicated ethically to pull off. The basic design is that members of a cohort of patients who have given informed consent are randomly assigned to either a treatment group or a "sham surgery" group. The treatment group gets the real operation. By contrast, patients in the control group are prepped for surgery and have a sham procedure, involving the drilling and sewing up of, say, a small hole in the skull. Critically important, the study is blind: the patients and the neurologists have no idea which group is which. This blinding controls for any placebo effect, which can be very large. One of the first sham surgery trials for a very common arthroscopic procedure for osteoarthritis of the knee found, surprisingly, that the sham surgery group reported the same benefit as the group getting the expertly performed operation. (And this particular arthroscopic procedure is no longer covered by Medicare.)

Such placebo concerns led two teams in the United States to propose controlled blind trials of fetal tissue transplant operations. Patients entering the trial would be informed that they would be assigned to one of two groups: a treatment group, which would receive the real operation implanting fetal grafts, and a control group, which would be prepped for surgery and have a small hole drilled in their skulls (but no grafts would actually be put in). Patients would have no idea which group they were in and would continue taking their regular dopaminergic medication.

One research trial, headed by the Denver neurosurgeon Curt Freed and the Columbia neuroscientist Stanley Fahn, randomly assigned Parkinson's patients under sixty to either a treatment or a sham surgery group and did the same for a group of Parkinson's patients over sixty. The doctors held follow-up appointments with the patients in the four groups for twelve months. In 2001, the team reported the results. The over-sixty treatment group experienced no measurable improvement compared with the placebo patients. The under-sixty treatment group did get some improvement not seen in the placebo controls, but the value of this benefit was called into question when investigators found worrying evidence of adverse side effects: facial dystonias and dyskinesias. Unlike L-dopa-induced dyskinesias, which disappear as patients' regular L-dopa medication wears off, these dyskinesias were coming from the graft, and they were permanent. The investigator Stanley Fahn calls these "runaway dyskinesias." He uses that term "because they were unstoppable. Eventually, we stopped L-dopa completely, and they're still having dyskinesias."*

A second study based in Florida involving the neuroscientist Warren Olanow, the neurosurgeon Thomas Freeman, and the neuropathologist Jeff Kordower was published in 2003. In this trial, patients were randomly assigned to two treatment groups—one using material from a single fetus, one using material from four fetuses—and a sham surgery group. The groups were followed for two years and evaluated using the UPDRS. There was no difference between the three groups, indicating that fetal dose didn't matter, and again some patients developed graft-related dyskinesias.

These two studies effectively killed the field of neural transplantation. The bottom-line conclusion of both: fetal surgery doesn't work consistently and carries significant risks. And Stan Fahn still thinks that given the risks and underwhelming benefits this was appropriate: "To justify all the risk of a transplant, it's got to really and dramatically be better than other alternatives." Kordower is also resigned to the failure, saying we have to accept what science is revealing to us. "You do an open-label study, patients get better; you do a blinded study, they do not get better. The only thing that correlates with better function is whether it is blinded or not."

A decade after the two studies, I visited Patrik Brundin in his new home in Grand Rapids, Michigan. The scrupulously polite Brundin, a lanky, youthful fifty-year-old, had recently been appointed assistant director of the Van Andel Institute, where he

---

*Paul E. Greene, a neurologist at Columbia University's College of Physicians and Surgeons and a researcher in the study, felt especially concerned about the dyskinesias. He wrote in the March 8, 2001, issue of *The New England Journal of Medicine* that some patients "chew constantly, their fingers go up and down, their wrists flex and distend. And the patients writhe and twist, jerk their heads, fling their arms about."

was setting up a world-class Parkinson's research center, still determined as always to find a cure for the disease that claimed his father.

Despite moving to the United States, Brundin is still unhappy about the American rejection of the neural grafting research. As he told me, the year 2001 was "a terrible time for cell therapy . . . a very political climate. George Bush had just become president, and using fetal tissue for anything or using embryonic stem cells was extremely negative . . . then these two trials literally hit us in the head. The field more or less came to a halt."

He is still a strong supporter of neural grafting. As he explains, "I still say yes, yes, they work when you do it right, and the problem is there are so many ways to do it wrong." He claims the two studies had numerous flaws. In Freed and Fahn's Denver surgeries, the tissue was not fresh but cultured for up to four weeks. The two teams followed different procedures from those used in Lund. The Denver study used no postoperative immunosuppressant at all. The Florida trial used only the immunosuppressant cyclosporine (as opposed to three drugs) for just six months, so thereafter the grafted tissue was exposed to the recipient's full immune system. The Florida study focused on the UPDRS motor section as an outcome measure, which critics consider flawed and insensitive. The Denver study used as its primary outcome parameter a subjective global rating scale, whereby the patients self-reported how they were feeling. Such an instrument is highly susceptible to placebo effects. The follow-up period in both trials, says Brundin, was simply too short. Swedish and British open-label trials suggested that improvements took much longer than twelve or twenty-four months to reach their maximum.

Brundin may turn out to be correct about neural grafting's efficacy. There's little doubt that it worked dramatically in the cases of two patients, who are known in the scientific literature as patient 7 and patient 15 in the Lund series. The two British men were diagnosed nearly thirty years ago. Both responded well to L-dopa but developed severe motor fluctuations and dyskinesias. So, in the 1990s, the two patients traveled to Sweden, where the surgeon transplanted dopaminergic fetal tissue into the striatum on both sides of their brains.

Ever since, Brundin says, doctors in the National Hospital for Neurology and Neurosurgery, at Queen Square in London, have carefully followed up patient 7 and patient 15. The principal researcher, Zinovia Kefalopoulou, and her colleagues report that for a long time the patients experienced little benefit. But after roughly four years, the positive effects had become unmistakable. First, both patients were able to completely drop *all* dopaminergic medication. Second, PET scans of the patients' brains show clear signs of new dopamine production in the striatum. Third, their motor states, as measured by the (admittedly limited) UPDRS, showed a sustained benefit. After four years, patient 7 had a score of 22, and patient 15, a score of 18 (out of a total of 108). These low scores—which indicate only mild impairment— remain unchanged to this day. For comparison, my UPDRS score on L-dopa is currently around 18.

Videos chronicling the progress of the patients are very convincing. A recent video of patient 15 shows him standing up from a chair, effortlessly walking down the corridor with good arm swings, and easily passing a balance test. Patient 7 isn't quite so impressive. While he moves fluently, he shows persistent dyskinesias from his neural grafts.

Even if it's just two patients, it's exciting to me. Such cases show that this bold strategy *can* work and serve as a caution against dismissing neural grafts prematurely. The fact that two controlled trials of neural grafting failed does not prove that this particular therapeutic strategy is hopeless. As the history of medicine reveals, if new therapies don't get the *details* right— the quality control, the appropriate patient cohort, the correct dose, the right response time—breakthrough treatments may be discarded. The early scientific trials of levodopa—*the* major achievement in the early history of Parkinson's research—failed every bit as much as fetal surgery. As we know, in L-dopa's first controlled double-blind trial in 1966, for example, the Swedish neurologist Clas Fehling found that levodopa not only had no effect on Parkinson's symptoms but also caused high blood pressure and nausea in a third of patients. But scientists like George Cotzias persisted and worked out the correct dosing regimen, and failure turned into spectacular success. The University of Rochester's Karl Kieburtz thinks that the same difficulties may be standing in the way of neural transplants. But, he says, "science is going to figure these things out."

And indeed, neural transplants may yet prove their value. In Europe, a large trial called TRANSEURO is under way involving some 150 patients in the United Kingdom, Sweden, and Germany. Unlike with the controlled trials in the United States, the investigators plan to give the transplanted cells time to mature— at least two years—and then follow up the patients for at least five years. If benefits endure well beyond what could be explained with a placebo effect—and few scientists think a placebo effect can last for five years—and if TRANSEURO avoids "runaway dyskinesias," the work might redeem neural grafting.

Further adding to the potential of this disease-modifying approach, in the last few years a potential alternative to fetal cells (and embryonic stem cells, which are equally controversial) has become available. In 2006, the Japanese researcher Shinya Yamanaka demonstrated in mice that ordinary skin cells could be reprogrammed, by manipulating just four genes, to become pluripotent—capable of becoming any cell. Soon after, Yamanaka's trick was achieved with human skin cells. This was, in principle, a major advance for diseases like Parkinson's. For now, rather than using fetal cells or embryonic stem cells, researchers can take a patient's own skin cells, reprogram them to become so-called induced pluripotent stem cells (iPSCs), then let them develop into dopamine neurons. These neurons can be studied in the laboratory or grown for use in neural grafts. Such iPSCs not only bypass the ethical issues plaguing embryonic stem cells, which have been a big drag on research for more than a decade, but also have other advantages: because iPSCs are derived from the patient's own cells, there is no need for immunosuppressive drugs. But because there is a risk that such cells might turn cancerous, developing a workable system that is safe and effective may take decades.

So while neural grafts may eventually play a role in the future treatment of neurodegenerative disease, the initial promise has so far not borne fruit. Luckily, during the last two decades, scientists have been hard at work on a second disease-modifying strategy. If you can't yet replace dead neurons with cellular grafts, perhaps you can protect surviving nerve cells and even revive them before they expire.

9

NEUROPROTECTION

By the time an "at-term" baby is born, her brain possesses approximately 100 billion neurons. Among those nerve cells are dopamine neurons in the substantia nigra. We depend on such dopamine nerve cells for numerous activities and processes, including movement, motivation, punishment, reward, cognition, mood, attention, memory, and sleep. Yet, remarkably, it turns out that only a tiny fraction of the 100 billion nerve cells—about 400,000—make dopamine. How can so few dopamine neurons do so many things?

The answer, researchers argue, is their prodigious interconnectivity. A dopamine neuron has the capacity to sprout massive numbers of branches along its axon—the fiber that carries the neuron's output signal. Scientists have estimated that a single dopamine neuron's axon can generate a dense forest of branches yielding more than one million synaptic connections, with a combined length of over four meters (a little over thirteen feet). While this enables the neurons to link up with many other brain cells and modulate numerous complex biochemical pathways, there's a cost to this interconnectivity: dopamine neurons require far more energy than other nerve cells. This means that these cells depend critically on their subcellular power stations—structures called mitochondria—to pump out the necessary energy to support their massive network activity.

That energy dependence in turn makes them especially vulnerable. Should enough power stations fail for any reason, those nerve cells will suffer distress and die. And that's just what happens. Every year, an estimated average of twenty-four hundred dopamine neurons die and can never be regenerated. The calculus is stark. About half of a healthy adult's lifetime supply of dopamine cells are dead by the age of eighty. There's a scientific consensus that once about 70 percent of dopamine-making cells in the substantia nigra die, the clinical symptoms of Parkinson's kick in. If everybody lived to 120, then everybody would hit this threshold just by aging. But some individuals hit the threshold earlier and manifest symptoms—many, like me, around sixty years old and some much earlier. In this case, other sources of cell death—genetic mutations or environmental toxins, for example—are likely involved in the decline in the reservoir of dopamine-producing neurons. And as the dopamine cells keep dying, the symptoms associated with Parkinson's get worse and worse.

Researchers have wondered why humans are the only species to get Parkinson's. Scientists argue that it's in part due to the fact that all animals—except the giant turtle—have shorter life spans, too short to develop the disease. But it may also be because many other species have proportionally more dopamine neurons than we do. A mouse weighing just a few ounces has 20,000 dopamine neurons, a rat 45,000. But a human, weighing around 175 pounds—hundreds of times heavier than a rodent—must make do with a mere 400,000 dopamine neurons (only about twenty times as many as a mouse).

Given that dopamine neurons are clearly a precious resource, how can we stop them from dying? It's a question of interest not only to researchers but also to people like me. If my disease could be slowed or stopped and I could hold on to my surviving dopamine neurons, then my prospects would be greatly improved. I could more easily accept my current state of impairment and worry less about my future decline.

Because dopamine neurons get "sick" for many reasons, researchers have come up with a number of neuroprotective strategies. One idea proposed back in 1985 was to protect dopamine neurons from the toxic effects of dopamine itself. Paradoxically, dopamine molecules can end up damaging the very neurons in the substantia nigra that created them. This is how scientists think it works: Neurons communicate by a process that is part electrical, part chemical. The signal that passes along the axon is electrical. But when the electrical signal reaches the end of the axon, this triggers the release of neurotransmitter molecules stored in tiny packages called vesicles. The vesicles then cross the so-called synaptic space that lies between the nerve terminal and the receiving portion of the next nerve cell.

Having achieved its goal of communicating, the cell needs to rapidly clear any leftover neurotransmitter out of the synaptic space so it doesn't interfere with future communications between the two neurons. Nerve cells accomplish this housekeeping in various ways. Some of the neurotransmitter is reabsorbed by the membrane of the first cell, in a process called reuptake. The rest is broken down by "janitor enzymes" into inactive small molecules. One special class of janitor enzymes that deactivates common neurotransmitters like dopamine and norepinephrine is known as monoamine oxidase, or MAO.*

The MAO degrades molecules of dopamine sitting in the synaptic space, yielding hydrogen peroxide and free radicals. Free

---

*One of the first classes of antidepressant drugs had been based on inhibiting the action of monoamine oxidase. Reasoning that depression resulted from a deficient supply of one or the other of the main monoamines, scientists had set out to keep what monoamines there were in the synapse and to prevent their reabsorption and deactivation. Because these early antidepressants directly attacked the metabolism of monoamines by MAO, they were called MAO inhibitors.

radicals—molecular fragments that have one or more unpaired electrons and have net negative charge—are bad news. They can wreak havoc in the human brain, damaging and destroying brain cells—including those in the substantia nigra. Some scientists even believe that free radicals are directly involved in the aging process itself. For these reasons, many health magazines and vitamin companies aggressively promote scavengers, such as vitamin E, which reportedly mop up free radicals.

Scientists reasoned as follows. If MAO degrades dopamine, producing hydrogen peroxide and associated free radicals, which kill neurons in the substantia nigra, then blocking MAO might be neuroprotective. In other words, treating patients with so-called MAO inhibitors early in their Parkinson's might slow the progress of the disease.

Interestingly, it was Walther Birkmayer, the Viennese physician who had first used L-dopa to treat Parkinson's disease in 1961, who first suggested this approach. In 1985, he conducted a study comparing a control group of 377 Parkinson's patients (on L-dopa alone) with 594 patients who received L-dopa plus an MAO inhibitor called selegiline over a nine-year period. And he found that the selegiline group lived on average fifteen months longer than the control group. Birkmayer's team interpreted these findings as evidence that selegiline was actually preventing the death of substantia nigra neurons in Parkinson's disease. But critics argued there was an alternative explanation: selegiline was acting symptomatically, producing a transient symptomatic day-to-day L-dopa-type effect in the patients. Such a symptomatic effect could look like a disease-modifying process was going on, because, on a day-to-day basis, patients would look as if they were doing better and progressing more slowly than the control group not taking selegiline.

Over the last two decades, neuroscientists and clinicians have built on Birkmayer's foundation and conducted a series of large, expensive, and highly technical placebo-controlled studies to rigorously test if selegiline and other MAO inhibitors (such as rasa-giline) are truly neuroprotective. And while there is no doubt that these drugs provide symptomatic benefit, the controlled trials of selegiline and rasagiline have failed to provide definitive evidence that they can protect neurons and slow the progression of Parkinson's disease. The dream of slowing the disease is not yet a reality.

But as there are many different ways that neurons become damaged, scientists have proposed other potentially neuroprotective therapies designed to block various disease pathways. Some researchers seek molecular targets that might protect or assist the mitochondria (the cells' power stations), which in turn might help dopamine neurons survive longer. Others argue that the neurons most vulnerable to degeneration have a limited ability to store calcium (as scientists have discovered that calcium plays a crucial role in the electrical behavior of neuronal firing). James Surmeier of Northwestern University has noted that substantia nigra dopaminergic neurons have a "pacemaker" function similar to heart cells; in the case of dopamine neurons, this characteristic helps ensure dopamine gets to the right places in the brain on schedule. The electrical mechanism driving the pacemaker involves the activity of calcium channels in the dopamine neurons' membranes. Surmeier's research on rodents suggests that excess calcium produced in these channels damages neurons and that *blocking* those calcium channels—so that the dopamine neurons switch to using more benign sodium channels instead—helps protect the dopamine nerve cells.

Interestingly, it turns out that one type of medication widely used to lower blood pressure also works to block calcium channels. The UCLA epidemiologist Beate Ritz analyzed a Danish population registry that tracked a large number of middle-aged people and found that those individuals who used a certain class of "calcium channel blocker" blood pressure medications—drugs that could, incidentally, cross the blood-brain barrier—had a roughly one-third reduced risk of developing Parkinson's disease. One such drug, Isradipine, is about to enter phase 3 trials as a potentially neuroprotective drug.*

These attempts to protect neurons have been somewhat eclipsed by another disease-modifying strategy, one that when I heard about it made me very excited. This approach focused on those neurons that were damaged but not yet dead. It asked the question: Can such flagging nerve cells be rescued and restored to health? It involves a class of brain chemicals called growth factors. And it is a gripping story.

In 1991, Frank Collins and Leu-Fen H. Lin, two scientists at the biotech company Synergen, based in Boulder, Colorado, isolated a brain protein that appeared to nourish and protect dopamine neurons. Because the protein was made by the glia (the cells that surround and insulate neurons), it was called glial-cell-line-derived neurotrophic factor, or GDNF. Collins and his colleagues produced a synthetic form of GDNF, and researchers at the University of Rochester and the University of Kentucky tested the novel brain fertilizer on dopamine neurons in test tubes and on monkeys rendered parkinsonian with MPTP. In the test tubes, the growth factor turned sick neurons into healthy ones. In the monkeys, the GDNF significantly reduced their parkinsonian symptoms. When executives in the California-based biotech company Amgen saw videos of the monkeys before and after administering GDNF, they were so impressed that they bought Synergen for $240 million and began planning for human trials.

---

*Clinical trials are conducted in a series of steps, called phases. In phase 1, investigators evaluate a new drug or treatment in a small group of people for safety, dosage, and side effects. During phase 2, the drug or treatment is tested on a larger group of people to continue safety evaluation and test for efficacy. During phase 3, the drug or treatment is given to large groups of people to further evaluate its effectiveness and monitor side effects. Phase 4 studies are done after the drug or treatment has been marketed to look for any side effects associated with long-term use or in special populations.

For Amgen, a lot was at stake. If sick neurons could be revived in people as well as in test tubes and experimental animals, then in theory Parkinson's disease might be slowed, halted, and even reversed. But this was no oral medicine like L-dopa. The principal challenge Amgen faced was delivering the GDNF to the right place in the human subjects' brains, notably the striatum, where surviving dopamine axons with nerve terminals might be found. This meant drilling a hole through the skull, carefully inserting a catheter, and, when it was correctly positioned in the brain, pumping in GDNF.

Between 1996 and 1999, Amgen carried out trials on thirty-eight human subjects. The researchers didn't attempt to reach the striatum directly, because the commercially available brain catheter was simply too large; instead they settled for delivering the GDNF to an easier target—the lateral ventricle—a location from which the scientists hoped the brain fertilizer would be carried by the cerebrospinal fluid to the striatum. The trial was a bust. After eight months, investigators found that the GDNF had not produced any clinical benefits but had instead caused serious side effects in some of the patients, who complained of nausea, delusions, and chest pains. One of the trial patients subsequently died of unrelated causes, and Amgen commissioned the neuropathologist Jeff Kordower to conduct an autopsy. Kordower found no evidence that the GDNF was working. As he told me, "There was no regeneration, no trophic effects . . . nothing."

Shortly afterward, the British neurosurgeon Steven Gill attended a lecture about the failed Amgen trial and had an epiphany about why it hadn't worked. The GDNF, he concluded, had not made it to the striatum. It had not even made it into the ventricle. Most of it probably wasn't even in the brain. The problem, reasoned Gill, was the size and design of the catheter. "If you just put a catheter into the brain and put fluid into it, it will just reflux. That's of course what happened with the Amgen trial." What you need to do, says Gill, is first place the appropriately sized catheter in the target site you want to reach (the striatum) and also deliver the GDNF at *high pressure*. That way it won't reflux; instead, the catheter will drive the fluid out a large distance so that it has a chance of reaching its target.

Gill set about designing his own mini-catheter and asked Amgen if he could have some of the remaining GDNF for his own research. Having abandoned the drug by that point, Amgen agreed. Gill mounted an in-house open-label study at the Frenchay Hospital in Bristol involving five moderately advanced Parkinson's patients. He surgically inserted two mini-catheters (one for each hemisphere) in the putamen, the section of the striatum most involved in motor coordination. Each catheter was fed by a pump, placed in the abdomen, that delivered a precise dose of the critical growth factor GDNF. Patients made a monthly trip to Bristol so Gill could renew the pump's reserve of GDNF.

According to Gill, all the patients showed improvements in their "off" states. After one year, says Gill, "all five patients showed dramatic changes . . . people moved better, their depression went away." Gill added provocatively, "They even became smarter." As supporting evidence, he proudly showed me videotapes of the five patients taken before and after treatment. Before the infusion, one of the patients took about five minutes to stand up and traverse a room. After infusion, he jumps out of his chair and walks briskly back and forth. The clinical benefits on the neurologists' rating scale, the UPDRS—which showed improvements of about one-third—were supported by PET scans, which demonstrated enhanced dopamine metabolism. There were no serious side effects.

Impressive though it looked, it was an uncontrolled study, an open-label trial—just like the ones done for neural grafting. The placebo effect is known to be powerful in surgical interventions and might well explain the dramatic improvements I'd seen on the videotape. As Kordower says, "If you're going to go through the rigors of surgery, you [the patient] really hope it's going to work. And the evaluators really hope it's going to work. And so you end up with a placebo effect."

For all its limitations, Gill's small, uncontrolled study rekindled Amgen's interest in GDNF, and the company carried out new animal studies and mounted a blind placebo-controlled trial with thirty-four patients—seventeen in a GDNF treatment arm and seventeen in a placebo control group. Unfortunately, Mickey Traub, the Amgen neurologist, could not use Gill's specially designed mini-catheter because it wasn't FDA approved in the United States (and FDA approval is a long and complex process). Instead, Amgen collaborated with Medtronic, a medical device manufacturer, which had a catheter division with instruments the team believed could do the job.

By 2004, it was clear that Amgen's second trial had failed. Patients receiving GDNF did no better than those in the placebo group. According to Gill, Amgen and Medtronic largely repeated the mistakes of the first trial. Even though this time Amgen attempted to infuse GDNF into the putamen rather than the ventricles, Medtronic's catheter was still too big and did not deliver the GDNF under pressure. Says Gill, "At the flow rates they were using, all that happened was liquid refluxing." More troubling for Amgen were safety issues. Brain autopsies from the monkeys showed a handful of them had toxic damage to the cerebellum. Amgen also reported that it had found neutralizing antibodies to the synthetic GDNF in two human subjects, raising concerns that these antibodies might diminish naturally occurring GDNF.

Around this time, Mickey Traub died of a heart attack, which signaled the beginning of the end for Amgen's ambitions with GDNF. Without his influence, things went from bad to worse. On September 1, 2004, Amgen announced it was halting all clinical use of GDNF around the world, a decision endorsed by the FDA in the United States and by the corresponding agencies in Canada (Health Canada) and the United Kingdom (the Medicines and Healthcare Products Regulatory Agency). This meant that trial patients in the United States, Canada, and the United Kingdom could no longer continue using GDNF. In the spring of 2005, some of those patients sued Amgen in two lawsuits, one in New York, the other in Kentucky. In both cases, the court ruled that Amgen had no legal requirement to continue distributing GDNF to the trial patients.

Had it all been a placebo effect? Gill, who is regarded as something of a maverick in the world of neuroscience, doesn't think so. Here's why. In 2005, one of Gill's patients died of a heart attack. Unlike the other four patients, whom he had trans-fused bilaterally, into both sides of the brain, this patient had received an infusion only on one side. Gill realized that an au-topsy could show a comparison of the GDNF-infused hemisphere and the untreated one. And he reported there was indeed a dif-ference. The infused side showed axonal sprouting in the pa-tient's putamen, where the catheter had been placed. This axonal sprouting was a sign that the GDNF had indeed nourished dopa-minergic neurons (although it is not clear if the nerve fibers made functional connections with receptors in the striatum). More-over, the untreated hemisphere showed no neuronal growth, only typical Parkinson's neuropathology.

To this day, Gill believes that his approach worked and that it offers a potential way forward. One of his most ardent support-ers is the Cure Parkinson's Trust cofounder Tom Isaacs. During his epic fund-raising walk circumnavigating Britain, Isaacs had met with Gill (shortly after walking across the Severn Bridge that connects Wales and England). "When I spoke to him," Isaacs recalls, "I really could see it in his eyes that he genuinely believed that he could cure this condition . . . and it was hugely excit-ing . . . And I just realized that's where I wanted the money to go to." With funding from Cure Parkinson's Trust, Gill is currently doing a larger study with thirty-six patients with a state-of-the-art catheter that has been tested on large animals, such as pigs. This trial will compare the effects of GDNF delivered every four weeks to the putamen region of the brain for a nine-month period with a monthly placebo infusion that would be expected to have no brain-cell-restoring effect. This trial will allow Gill and his colleagues to determine if direct GDNF infusion has potential as "a disease-reversing therapy in Parkinson's disease."

But from the beginning, some neuroscientists argued that there was a better way to deliver growth factors to parkinsonian brains than by direct infusion—using genes.

Gene therapy is an intriguing concept. You start with a simple virus like a cold virus. You gut it of its own genes and replace them with the gene or genes of your choice—in this case, the gene that encodes the growth factor GDNF or its cousin, a closely related growth factor called neurturin (NTN). Then you need to fire multiple copies of this virus into the patient's putamen, where, if all goes well, instead of giving the brain a cold, the therapy infects neurons with the desired gene, and those neurons start expressing growth factor—the "fertilizer" with the potential to revive nearby dying cells and axons. Unlike infusion (where the growth factor needs to be continually renewed), gene therapy is a one-shot process. Once the genes are inserted and turned on, they should keep working indefinitely. Starting in 2000, researchers began pursuing this exciting but technically challenging option.

It started out well. In 2000, the ubiquitous Rush University scientist Jeff Kordower published a paper in *Science* demonstrating a proof of the concept in monkeys. He gave monkeys MPTP, the contaminant in the California street drug that had caused the addicts to freeze up, for between five and seven days. Then, rather than allowing the monkeys to become parkinsonian, he injected GDNF-gene-carrying viral particles into the striatum and substantia nigra. According to Kordower, "It protected virtually the entire nigrostriatal system and also enhanced function in aging monkeys. It rescued all of the neurons."

Based on this and other promising preclinical studies, Kordower became a founding member of a new San Diego–based biotech company called Ceregene Inc. Because many scientists want to see their work commercialized, it has become routine for academic researchers to participate in founding and advising fledgling biotech companies. With seed money from venture capitalists, Ceregene undertook a phase 1 safety trial in human beings. As Amgen held the patent on the gene for GDNF, Ceregene could not use GDNF but instead employed the gene for neurturin in the trial, giving it to twelve patients with advanced Parkinson's at two clinical trial sites—University of California, San Francisco, and Rush University Medical Center in Chicago. All twelve patients enrolled in the open-label study underwent stereotactic neurosurgery to deliver one of two neurturin dose levels into their putamen. After a year, patients' UPDRS motor scores had improved by around 40 percent. Patients also reported a reduction in "off" time, and clinicians noticed no serious adverse events.

But again, this study hadn't attempted to control for any placebo effect. So Ceregene followed up with a phase 2 double-blind placebo-controlled trial. Fifty-eight patients participated in the study, and the researchers randomly assigned thirty-eight of them to gene therapy treatment and the remainder to a sham surgery group, where they underwent a surgical procedure that only appeared authentic. Each patient in the treatment group received viral infusions at eight sites (in the putamen). The sham surgery control group received no infusions. The results, reported in 2008, were somewhat shocking. After one year (the agreed end point), there was no difference between the gene therapy and the placebo groups. In a controlled setting, there was simply no evidence that neurotrophic therapy worked better than a placebo.

Starting a biotech company is not for the faint of heart. If clinical success is not demonstrated within a decade, investment capital will likely dry up. Ceregene's CEO, Jeff Ostrove, considered shutting down the NTN gene therapy program, and the company was forced to lay off more than half its staff. But then two trial patients died of unrelated causes, providing an opportunity to examine their brains. The autopsies revealed that the infusion had fallen short of expectations. Only 15 percent of the putamen expressed the neurturin gene; Ceregene researchers had been hoping for a figure closer to 50 percent.

Just as L-dopa had initially failed because of inadequate dosage, researchers wondered if better dosing and delivery of NTN would change the odds for neurotrophic factors. With a $2.5 million grant from the Michael J. Fox Foundation, Ceregene launched another trial involving fifty-one patients, infusing four times the viral dose and delivering the NTN to the substantia nigra as well as the putamen.

In April 2013, the company announced the new results to patients, researchers, and investors eagerly awaiting news from the front lines. The news was bad. After a fifteen- to twenty-four-month follow-up, there was no difference between the group that received gene therapy and the placebo group. The intervention had failed.

Is this the end for growth factors? Proponents still hope that the Gill infusion study now under way in the United Kingdom and a new twenty-four-patient NIH gene therapy trial using GDNF will show enough promise to keep the research alive. But based on what I have learned, I am not very optimistic. To me, it seems more likely that growth factors will be eclipsed by new alternative therapies working their way to the clinic (which I will discuss soon).

Some fifteen years after Michael J. Fox had optimistically started his foundation, the three disease-modifying approaches—protecting, reviving, and replacing dopamine neurons—have faced setback after setback. Researchers continue investigating whether changing the key parameters related to dose, delivery method, patient cohort, and follow-up period will turn failure into success, as had happened with L-dopa in the 1960s. But the idea that such therapies would be ready in five years, as Senator Specter had suggested in 1999, now seemed ridiculous to those who had been at that Senate hearing.

How could scientists have been so wrong about these disease-modifying interventions? Researchers pointed to a number of methodological challenges that plagued clinical research in Parkinson's disease—from the pervasiveness of the placebo effect to a risk-averse pharmaceutical industry, from the limitations of the UPDRS to the masking effects of L-dopa medication.

But some scientists and clinicians argued the failures reflected something more fundamental: that the classic picture of Parkinson's disease as a dopamine-centered motor condition focused on one tiny brain region was out-of-date. Before researchers and pharmaceutical companies could deliver a cure, the Parkinson's community of neuroscientists, clinicians, and patients would have to negotiate a paradigm shift.

REBRANDING
PARKINSON'S DISEASE

Most therapeutic interventions against Parkinson's shared a weakness: they were targeted only at the dopamine system, and therefore they could mitigate only the classic motor symptoms identified by James Parkinson and Jean-Martin Charcot back in the nineteenth century. In reality, however, there was increasing epidemiological, clinical, and pathological evidence by the first decade of the twenty-first century that the classic motor symptoms of Parkinson's disease were only part of a more complex medical narrative, one in which the disease emerged long before any tremor became noticeable and developed later in life to include many symptoms that had nothing to do with movement—from constipation to sleep disorders, from double vision to dementia. If this evidence survived scrutiny, researchers would need to rethink their strategies for combating the disease.

Interestingly, throughout history there have been patients who clearly displayed this wide range of symptoms. Neurologists once considered them to have a distinct condition called Lewy body disease, named after Frederick Lewy, the pathologist who identified Lewy bodies in 1912. But today most neuroscientists consider their condition a compressed, "sped up" form of Parkinson's. As such, victims of Lewy body disease present a more complete

symptomatic picture of the multiple signs and symptoms a Parkinson's patient will confront over the course of his life. One such
afflicted individual whom I was privileged to meet is Thomas
Graboys. To me, his story dramatically illustrates that Parkinson's disease involves so much more than movement problems.

His personal narrative is like a Greek tragedy. In the mid-1990s,
Graboys's life was as good as it gets. Photographs of Graboys
show a man in his element: a lean, athletic physician adorned
with a white coat and stethoscope, looking into the camera with
a friendly, confident smile. Happily married with two daughters, a
star on the faculty of Harvard Medical School and the Brigham
and Women's Hospital, Graboys says, "I was on top of the world
in every way. I was blessed." Graboys even became a minor celebrity when the Celtics picked him as one of a team of Boston car-
diologists charged with caring for the basketball legend Reggie
Lewis after he collapsed during an April 1993 NBA practice game.*

But Graboys's charmed life began to unravel. In 1998, his wife,
Caroline, died of colon cancer. Then he began to notice a series
of small things go strangely amiss with his body. An avid tennis
player, he found his usually strong serve was going into the net.
Skiing in Vail, he lost his balance and fell over—something he
almost never did. He started to experience symptoms of rapid
eye movement (REM) sleep behavior disorder (RBD). Here, the
body's normal practice of paralyzing muscles during dreaming
or REM sleep breaks down, and the limbs shake. Someone experiencing RBD is prone to sleepwalk and may hurt himself or his
sleep partner by lashing out with involuntary movements. In Tom's
case, acting out his dreams during a sleepwalking episode resulted in a bad fall that cost him two of his teeth.

*Lewis, whom the dream team diagnosed with cardiomyopathy, collapsed again and subse-
quently died on July 27 of that year.

He soon encountered clear-cut evidence that something was really wrong. He developed a tremor and a stooped posture, and his problems with balance only increased. He suffered mental lapses and fainting spells. Graboys tried to keep his condition secret from colleagues, patients, friends, even family—including his second wife, Vicki Tenney, whom he married in 2002. But keeping health secrets in a medical town like Boston is all but impossible. In his memoir, *Life in the Balance*, Graboys relates his dramatic "outing." While he was walking to his car one day in 2003, in the parking lot underneath the Brigham and Women's Hospital, he explains, "a voice called to me from a short distance away. It was Martin Samuels, chairman of the Neurology Department at Harvard Medical School and Brigham and Women's Hospital. He knew nothing of the drama unfolding in my life, but he said to me, 'Tom, who is taking care of your Parkinson's?' I was stunned. I hadn't even been diagnosed with the disease . . . Yet Martin knew just from looking at me."

Graboys learned that he had a particularly aggressive form of Parkinson's, sometimes called "Parkinson's with Lewy body dementia." He confronted a more extreme version of what normally plays out for the average Parkinson's patient over fifteen to twenty years.

To better understand the full range of symptoms I was likely to confront at some time in the future, I visited Tom several times at his beautiful home in Chestnut Hill, Massachusetts. On the most recent occasion, after meeting me at the door, Tom led the way inside, taking short quick steps, as if his ankles were chained together, his body flexed forward like a hunchback's. Twice, he shuffled to a halt, his shoes seemingly glued to the floor. For a few seconds, he hovered there as frozen as a mannequin. Then the spell broke, his shoes magically came unstuck, and he continued on his way to a small lounge.

His voice was very, very soft while we spoke about his life with Parkinson's, and it took all my concentration to process

what he had to tell me. Tom veered between lucidity and episodes of cognitive breakdown. When that happened, he lost track
of the topic at hand and sat for minutes struggling to remember
what he'd been saying. Sometimes a little prod from me got him
going again. Other times, the particular mental schema had vanished, and he could not recall it. But mostly the old Tom—clever,
thoughtful, and compassionate—was still there.

Ever the physician, he inquired about *my* condition. "How is
your tremor?" he asked.

So far, my symptoms were holding steady, I explained to him.
Tom's news, on the other hand, was awful. He'd fallen multiple
times and been rushed to the Brigham and Women's emergency
room, and these setbacks had shaken his self-confidence. "I used
to think that people who were progressing fast were not trying,"
he told me, "that they were weak-minded individuals . . . I realize
now that isn't true. Over the last year, I've become fragile. It's
ultimately very demoralizing to realize that the disease is progressing, whatever I do."

Indeed, he and Vicki had just returned from a "demoralizing" trip to California, in which he'd gotten stuck in the airplane's
bathroom in an episode of gait freezing. Such experiences, he
said, make him reluctant to leave home. "My house is a safe haven . . . it's secure . . . I could stay in here pretty much forever."

Eleven years into the disease, he's surprised that he's still
alive. But impressively, in the midst of his decline, he's become an
ambassador of hope. "In the face of uncertainty, there's nothing
wrong with hope," he says.

I felt compassion for Tom and admired his courage. But
mostly I sensed the dark force taking over his body and mind. In
little more than a decade, he had progressed from living with
what I have—mild symptoms on one side of the body (so-called
stage 1 Parkinson's on the Hoehn and Yahr scale)—to being severely disabled but still able to walk (stage 4). When Tom reaches
stage 5, he'll be wheelchair-bound or bedridden. But I found

something else even more striking. Tom displayed not only Parkinson's classic motor symptoms but many other conditions as well. Tom had it all: from neuropsychiatric symptoms (depression, anxiety, hallucinations, and cognitive impairment) to sleep disorders (REM sleep behavior disorder, insomnia), from disabling fatigue and chronic pain to visual disturbances, from bladder issues to impotence, from constipation to sudden drops in blood pressure. As he'd told me, some of these symptoms had preceded the diagnosis of his Parkinson's disease by many years.

Tom was living evidence of what scientists were arguing: Parkinson's disease was definitely not just a movement disorder. This theory held that the traditional scientific narrative—that Parkinson's was all about the death of dopamine neurons in the substantia nigra—was an oversimplification. According to this new view, the dopamine-related symptoms were just the tip of a clinical iceberg. Because that clinical iceberg contained common maladies of the middle-aged and the elderly—constipation, sleep problems, depression, and the like—clinicians had only recently grasped the connection with Parkinson's disease.

Where did this view come from? In truth, some clinicians had long observed that regular Parkinson's patients had symptoms that seemed completely unrelated to movement. The three most commonly noted were problems with sleep, constipation, and smell. James Parkinson had mentioned two of them. Back in 1817, he had reported that "sleep becomes much disturbed. The tremulous motion of the limbs occur during sleep, and augment until they awaken the patient, and frequently with much agitation and alarm." He also noted constipation was a common symptom, saying that the bowels "which had been all along torpid, now, in most cases, demand stimulating medicines of very considerable power."

There was also historical evidence that pathologists, including Lewy, had noticed that the characteristic Lewy bodies did not just show up in the substantia nigra but were found in other regions of the central nervous system, the peripheral nervous system, and the gut. In 1975, the scientists K. A. Ansari and A. Johnson reported that extensive Lewy body pathology could be found in the olfactory bulb, the part of the brain that processes smells. But at the time, little was made of all this.

Then, in the 1990s, early reports started coming in from a long-term epidemiological study of a large cohort of Japanese American men in Hawaii. The primary cohort of 8,006 men, all born between 1900 and 1919, had originally been recruited for a Veterans Administration (VA) heart and stroke study, the Honolulu Heart Program, which conducted regular questionnaires about lifestyle (to assess smoking, diet, exercise, and more) and periodic clinical tests (to track blood pressure, EKG, blood lipids, blood sugar, and other metrics) on the men. In 1990, the VA and the NIH realized that the surviving 4,678 members of the cohort (now in their seventies and eighties) would be a perfect group in which to study neurodegenerative diseases like Parkinson's and Alzheimer's. So the VA recruited the neurologist George Webster Ross to run the so-called Honolulu-Asia Aging Study. Ross would have access to the surviving members of the cohort, which contained several hundred Parkinson's patients. Like neurologists before him, Ross found that many of these Parkinson's patients also had symptoms like constipation, impaired smell, and sleep problems. And then Ross and his colleagues came up with a profound idea. "We said wouldn't it be interesting to see if they had these symptoms *before* their diagnosis of Parkinson's disease?" Clinicians call signs and symptoms that predate a diagnosis of disease "prodromal," from the Greek word *prodromus* for "forerunner."

The remarkable thing about the Honolulu-Asia Aging Study was that it allowed researchers to do just that: to travel back in time (as far back as 1965) and look at data on behavior related to everything from smoking, alcohol, and coffee consumption to the number of daily bowel movements of the participants. Because these data had been obtained prospectively—*before* any individuals had contracted Parkinson's disease—Ross was able to search for factors that appeared to increase or decrease a person's risk of getting the condition.

What Ross and his colleagues found was striking. They looked first at constipation. "We found that individuals with less than one bowel movement every day were about three times more likely to get Parkinson's than those individuals who had more than one bowel movement a day. This could be detected at least twelve years before the diagnosis of Parkinson's." And, says Ross, there was a quantitative relationship as well. "So the more bowel movements an individual had, the less likely he was to get Parkinson's disease." The researchers also looked at smell and established scientifically that impaired smell was an early predictor of Parkinson's. Other studies found an association with sleep disorders. If an individual has three or more of the associated prodromal symptoms, says Ross, he or she is "about ten times as likely to get Parkinson's disease as somebody who doesn't have any of these features."

Interestingly, other factors—drinking coffee and smoking cigarettes, for example—seemed to *reduce* the risk of developing Parkinson's disease. Says Ross, "The protective effect of high caffeine and high smoking . . . is so strong that if you could combine them into one therapy it is reasonable to think it could be effective."

If Ross's work suggested that the disease is present long before a patient experienced motor symptoms like tremors and rigidity, it also showed the clinical picture at the end of life is somewhat more complicated than James Parkinson and Jean-Martin Charcot had described. Ross found that virtually all of the elderly survivors (those who survived into their late seventies and eighties) displayed cognitive deficits and dementia. Other epidemiological investigators found similar results. In the 1980s, the Australian scientist Mariese Hely began following one hun-

dred Parkinson's patients. In 2008, she reported that after twenty
years, 74 percent had died, and the vast majority of the survivors
had severe constipation, urinary problems, swallowing issues,
balance challenges, and dementia. In short, they ended up a lot
like Tom.

In the light of this evidence, many neuroscientists are lobby-
ing to rebrand Parkinson's disease from a motor disorder to a
whole-body condition, involving an enormous number of signs,
symptoms, and complaints. These include not only the classic
features—such as bradykinesia, tremor, rigidity, postural instability, stooped posture, shuffling gait, freezing of gait, dystonia,
facial masking, small handwriting, dysarthria (problems with
articulation), dysphagia (trouble swallowing), oily skin, bladder
problems, pain, constipation, and loss of smell—but a growing
list of other problems as well.

People with Parkinson's experience neuropsychiatric symptoms
such as depression, anxiety, hallucinations, cognitive impairment,
and impulse control disorder (the last caused by dopamine agonists). They suffer from a whole host of sleep-related disorders—
including REM sleep behavior disorder, excessive daytime
sleepiness, restless legs syndrome, insomnia, and disordered
breathing while sleeping. They live with many types of pain. They
experience fatigue. They display apathy. There's a breakdown of
the body's autonomic system, leading to sexual dysfunction, excessive sweating, and sudden drops in blood pressure. There are
gastrointestinal problems, from loss of taste to drooling. There's
weight loss. There are visual problems, from double vision to inflamed eyelids.

A typical patient won't, of course, have all of these. And I
suspect that many of these signs and symptoms may turn out to
have more to do with general aging than with Parkinson's disease. In my experience, many complaints are transient as well. I
went through a period in 2013, for example, where I suffered
from occasional double vision, and then this problem simply

vanished. The same thing happened with constipation. For a few months, it was a problem, and then suddenly it wasn't. I also found that a number of unpleasant symptoms—for example, extreme fatigue and excessive daytime sleepiness—disappear when I get an adequate amount of sleep, and that this in turn takes care of itself when I exercise regularly. These changes may reflect the brain's attempts to take over functions with alternative brain networks and circuits.

Toward the end of my visit with Tom, he showed me a family photograph: Vicki, himself, and five of his thirteen grandchildren. He complained that because of his dementia he couldn't remember all the kids' names. I reminded him of his wonderful, rich life, both personal and professional: a reality that couldn't be stolen. Indeed, this is something I have been reflecting on myself as I come to terms with my altered future. He then shared with me a short but powerful letter he'd published in the journal *Movement Disorders* about why, despite all his burdens, he still valued his life. The soul, he writes, "is where hope lives; not a naive hope that I will, by some miracle, have my former self restored, but hope that tomorrow, and the day after, can still be days from which a measure of joy and meaning can be derived. And from hope springs optimism that, even with great limitations, there is life to be lived."

I drove home in a somber mood. Meeting Tom had forced me to contemplate matters most of us avoid thinking about. People with Parkinson's often like to imagine our decline will be so slow that we'll die gracefully of old age before the disease gets us. We focus on outliers like Pamela Quinn who live well and live long. But Tom, I realized, might represent a more complete picture of the end of life for an individual with Parkinson's. As I drove, I wondered, woud I be as disabled as Tom is toward the end of my life? Although I hadn't wanted to confront it, I needed to address the issue straight on. If branding Parkinson's disease primarily as a movement disorder was simplistic, what should this view be replaced with? And what did this paradigm change imply for treatment and cure? As I would soon find out, the consequences are profound and have implications not just for Parkinson's research but for many chronic diseases.

If we think of Parkinson's disease as a timeline, we can imagine my disease starting silently about twenty years ago, when I was forty. For twenty years, a pathological process was under way, yielding odd symptoms (like my worsening sense of smell) that never got my attention. By the time I was sixty, the damage to my nigrostriatal system was so advanced that I began to notice motor symptoms (like a tremor in my left hand), which led me to visit that center in Portland where a neurologist made his diagnosis. And since I learned of my fate in January 2011, my disease continues to progress as the timeline stretches into the future. While clinicians treat the present (a patient's current symptoms), researchers can look backward in time to explore where the disease came from and forward in time to investigate the accumulated pathological damage at death.

Just what kind of underlying pathology can explain such a long timeline and such a wide range of symptoms? Clearly one that isn't confined to the dopamine system but involves many other brain and body regions as well. Clues would emerge from two major intersecting scientific breakthroughs in basic research. The first looked back to the start of the disease; it concerned an extraordinary discovery in genetics made by a team of American and Italian scientists, revealing a new "bad actor"—a rogue protein called alpha-synuclein. This genetic discovery in turn enabled a brilliant German pathologist, Heiko Braak, to undertake a heroic series of anatomical investigations of the brains and bodies of Parkinson's patients who had died (sometimes of unrelated causes) at different stages of the condition. His efforts generated a provocative theory of Parkinson's disease that tied together not only Parkinson's motor symptoms like tremor, stiffness, and slowness but also the host of non-motor symptoms that patients were reporting.

It all began with a clinical encounter in the 1980s, at the Robert Wood Johnson Medical School in New Brunswick, New Jersey, in the department headed by the legendary neuroscientist who had helped pioneer L-dopa therapy, Roger Duvoisin.

THE DESCENDANTS

Chance favors the prepared mind.   —Louis Pasteur, 1854

After he helped pioneer L-dopa therapy in the 1960s with individuals like patient 3, Roger Duvoisin went on to spend much of his career puzzling over the issue of whether Parkinson's disease was primarily environmental or genetic. Duvoisin's hunch was that environmental factors, such as pesticides and air and water pollution, caused Parkinson's disease, and he held this view even before MPTP was discovered in the frozen addicts' street drugs. While it was true that in some rare families a Parkinson's-like condition appeared to be inherited, the vast majority of regular Parkinson's cases—which neurologists called "sporadic" Parkinson's disease, to indicate that it occurred without a discernible pattern or specific cause—did not seem to be genetic in origin. In addition, a series of studies involving identical twins (one of which Duvoisin headed)—the gold standard for detecting heritable traits—had failed to find evidence that Parkinson's had a clear genetic component.

So the discovery of MPTP in the 1980s seemed to clinch the matter for Duvoisin. As he recalls, "That seemed to say there was no genetics involved."

Then something happened that made him change his mind.

It all began with a routine office visit. In the spring of 1986 (the year my *Nova* documentary was broadcast), the neurologist Larry Golbe—Duvoisin's junior colleague at the Robert Wood Johnson Medical School—conducted a clinical examination of a forty-eight-year-old northern New Jersey fire chief named David. Golbe observed that the fire chief's movements were slow and restrained. Golbe had asked David to participate in a finger-tapping exercise (a common way for researchers to detect abnormal movements), and although David started the exercise with no trouble, he soon ran out of energy, and his finger taps got smaller and eventually petered out. When he stood up, this once athletic man now bent forward, with a stooped gait. When he walked, he didn't swing his arms, but shuffled along with small steps. When Golbe tried to bend David's arms and legs at the elbow and the knee, he was met with resistance. David's face was expressionless, and he never blinked.

Golbe flipped through the chart, scanning his patient's medical history. Ten years earlier, at the age of thirty-eight, the fire chief had reported stiffness and slowness in his left hand, and soon after he was indeed diagnosed with Parkinson's. David was treated with Sinemet, the carbidopa-levodopa combination that replaced some of his missing dopamine. For a while, the chief responded. But over time the medicine became less effective, as we've seen in the cases of countless people with Parkinson's, including Joan Samuelson, Michael J. Fox, and Tom Graboys. After a decade of drug therapy, David needed to take Sinemet every two hours to avoid freezing up. And even then, the medicine's effects sometimes suddenly switched off, and he became rigid as a statue. David also showed clear signs of dementia while he was "off," as well as postural instability, and manifested a festinating gait when he was "on." Golbe enrolled David in a new study that he was conducting of deprenyl (also known as selegi-

line, the drug we met in chapter 9), which neuroscientists hoped
would protect the neurons from further damage. But so far David
had reported only minor benefits.

Golbe—a dedicated, compassionate clinician whose own
father had contracted Parkinson's disease (as was the case with
the Swedish neuroscientist Patrik Brundin)—unfortunately had
little else to offer the fire chief. So he spent a few minutes counsel-
ing David and made a follow-up appointment with him for three
months later. That was the last Golbe saw of him. A few weeks
after their meeting, David tragically drowned in a swimming pool.

After the funeral, David's brother Frank came to see Golbe
seeking his own medical care, as he was concerned that he also
might have Parkinson's. After giving Frank a full examination,
Golbe confirmed that he did in fact have the disease. This got
Golbe's attention. As he had been trained to do by his boss and
mentor Roger Duvoisin, Golbe initiated a broad family study
to search for any other relatives who might have contracted
Parkinson's—brothers, sisters, aunts, uncles, cousins, nephews, and
nieces—and eventually unearthed a total of six related individuals
with classic parkinsonian symptoms. During his examinations
of the family members, Golbe recalls that the patients told him
"the family originated in Contursi, Italy, and that David's grand-
parents had emigrated to the U.S. around 1900." He didn't know
at the time if this was significant or not.

Several months after David's death, Golbe got a visit from a
Staten Island woman who presented with classic symptoms of
Parkinson's. After Golbe had examined her, he wondered whether
there might be something wrong apart from the Parkinson's disease. The patient, Joyce, had a Scandinavian last name, and this
made Golbe suspect she could also be suffering from pernicious
anemia (which is more common in that part of the world than in
the United States). But Joyce told Golbe that this was her husband's
family name and that she was of Italian descent. Specifically, she

told Golbe, her family came from a small village in southern Italy: a village in the hills of Salerno province called Contursi.

Chance favors the prepared mind, and Golbe immediately made the connection between David and Joyce. He realized he might have stumbled on a rare family "kindred," one that passed Parkinson's disease from generation to generation, a kindred geographically centered on Contursi, Italy. He called Duvoisin, and together they embarked on a complex international task of medical detection. Because of the connection to Contursi, Golbe realized he needed an Italian partner, so he looked in *The Italian Journal of Neurological Sciences* for articles by authors based in Naples, the city nearest to Contursi. He wrote to one, Dr. Vincenzo Bonavita, inviting him to collaborate. Bonavita passed the task to his colleague Giuseppe Di Iorio, who had a background in neuro-genetics. And so it began.

A year later, Larry Golbe was sitting in the small office of Dr. Salvatore La Sala in Contursi. Dr. La Sala, who had grown up in the village, was one of its only resident primary care physicians— and also served as its dentist. Golbe watched as his Italian collaborator, the neurologist Giuseppe Di Iorio, from the University of Naples, conducted a clinical examination of a forty-year-old man named Mario. Periodically, Dr. La Sala spoke to clarify Di Iorio's requests and reassure his fellow Contursian. Golbe wished he understood more Italian. Despite taking language lessons, he could barely follow the conversation. Di Iorio's English was also rudimentary, but somehow they managed to work together effectively, eventually cracking a genetic mystery.

After Golbe had plotted the American branches of the family tree, he suggested that Di Iorio visit the village church and examine Contursi baptismal and marriage records going back twelve generations. When they plotted the family tree on a huge chart, the multigenerational "pedigree" showed that Golbe's patients David and Joyce were related—they were seventh cousins. They were two of 574 descendants of a couple who married around

1700. Others, the team discovered, now lived in Italy, Germany, Argentina, Canada, and the United States. The truly remarkable finding, however, was that 61 of the recent descendants had developed Parkinson's disease. The pedigree analysis showed that males and females were equally affected and that descendants had a 50 percent chance of contracting the bad gene and, along with it, Parkinson's disease. In geneticists' parlance, the pattern of inheritance was "autosomal dominant."

Despite his limited Italian, Golbe followed Dr. Di Iorio's clinical examination of Mario with little difficulty. Based on his clinical signs, Mario had inherited the bad gene and along with it Parkinson's. Dr. La Sala, interestingly enough, was also a member of the family but hadn't inherited the mutation. He had played a key role mediating between the scientists and the family, explaining, for example, why the team needed to collect blood samples to take back to New Jersey for DNA analysis. Such molecular investigations might identify the specific genetic mutation and provide clues as to how it caused Parkinson's to develop in the bodies and brains of affected members of the kindred.

Meanwhile, back in New Jersey, other members of Duvoisin's department discovered a critical missing piece of the puzzle: they were able to confirm that the kindred members had genuine Parkinson's disease with the telltale pathology. They had obtained and examined autopsy materials from two deceased family members—David, the fire chief, and his maternal uncle. Their brains showed extensive damage to the substantia nigra, and some of the surviving dopamine neurons contained Lewy bodies—the blood-cell-sized structures we met earlier in the book and the pathological hallmark of true Parkinson's disease. As Duvoisin says, "It was classic Parkinson's disease pathology"— the first family kindred "where there was autopsy confirmation that it was Parkinson's disease."

Duvoisin realized that scientists, himself included, had been too quick to dismiss the idea that Parkinson's could be inherited.

The New Jersey team and their Italian collaborators had not only proved beyond doubt the importance of genetics but also unearthed a brand-new clue to the nature of Parkinson's. The next step was to find the mutant gene that caused one in two children in this family on average to contract Parkinson's, for that gene might hold the key to the mystery of the disease.

Unfortunately, in the years ahead the New Jersey team failed to capitalize on its discovery. The group of scientists appears to have lacked the specialized molecular skills to complete the next step. Years later, the team had made little progress toward finding the gene responsible for the Contursi disease kindred. And other scientists—who didn't have access to the unique set of blood samples—were getting restless. Matters would come to a head in 1995, at a meeting hosted by the National Institute of Neurological Disorders and Stroke in Washington, D.C.

On August 28, 1995, some two dozen top scientists from around the world and Parkinson's advocates converged on the Madison Hotel in Washington, D.C., for a special NINDS workshop about the malady. Over the next two days, they would discuss all the hot major research areas—from fetal tissue brain transplants to new drug therapies that sought to protect neurons from environmental toxins. But the NINDS director, Zach Hall, who had organized the workshop, believed that the most important contribution of the conference might come from the latest findings in the field of genetics, one of the less fashionable Parkinson's research areas at the time.

To promote the importance of genetic studies, Hall had invited Roger Duvoisin to present a progress report on the Contursi kindred. Duvoisin addressed the assembled scientists sitting around a large U-shaped table. In the discussion that followed, it became clear that while some scientists were still skeptical about the importance of genetics, others were frustrated at the slow progress being made toward finding a gene. It had now been more than seven years since Golbe and Di Iorio had completed the family pedigree, and people were becoming impatient at the lack of progress. At one point, Stanley Fahn, the influential clinician and researcher from Columbia University, directly confronted Duvoisin about the delay. Fahn recalls saying, "Roger, you've had years to find the gene. Don't you have an obligation to share this kindred with the scientific community?" Duvoisin says he replied, "Of course we will share it; we want the analysis done too."

Zach Hall recalls thinking that a great deal was at stake. The Contursi kindred was the first ever confirmed case of inherited Parkinson's disease. It was crucially important, Hall believed, for scientists to find the gene quickly, because it might hold the clue to understanding and perhaps curing Parkinson's. The problem was that gene hunting required state-of-the-art techniques being developed by the recently launched Human Genome Project, the initiative to map and sequence all the human genes. And based on everything he had heard, the New Jersey group seemed to be weak in such molecular techniques. According to Hall, "They had done this beautiful job of tracking down all these people, all over the world, clinically assessing them, constructing the pedigrees, I mean it was just a wonderful clinical detective job. And they knew there was a possibility of finding the gene, but they didn't know how to do it. It was clear they were in over their heads."

On the other hand, the NIH employed people with the skills to pull off feats of molecular wizardry, people like Bob Nussbaum, a forty-six-year-old clinical geneticist with advanced molecular biology training. After the meeting, Hall asked Nussbaum if he would be interested in mapping and sequencing the gene if the NIH could negotiate a collaboration with the New Jersey group. Nussbaum was enthusiastic about the idea and suggested that he work with his colleague Mihael Polymeropoulos.

Hall called Duvoisin to ask him to send the Contursi blood samples to the NIH so Nussbaum and Polymeropoulos could begin their analysis. According to Hall, initially Duvoisin argued that his team could do the job. Hall disagreed. "So we had to actively persuade them and so . . . we used a little bit of a carrot and stick, saying, . . . We can put you in touch with the best molecular people in the country and you can all share the glory . . . [but] you have an obligation to let this go ahead." Hall also agreed to kick in some funding so that the Robert Wood Johnson scientists could revisit Contursi to collect additional blood samples for genetic analysis.

In early 1996, the New Jersey group delivered to the NIH a set of blood specimens collected from living members of the kindred. After seven years of little progress, Nussbaum and Polymeropoulos now racked up some spectacular advances.

To find the rough area where the gene is located—the gene locus—geneticists use a process called linkage analysis. Researchers use known locations along the genome—so-called genetic markers—as chromosomal reference points. Then they track these genetic flag posts through time, looking for the flags that are always there in the DNA of people who inherit Parkinson's and never there in the DNA of those who don't. A marker that crops up only when the disease is present is said to be "linked," and the supposition is that the marker and the disease gene share a small region of the genome, because they are always inherited together. By taking blood samples from large numbers of both affected family members and healthy relatives over several generations and extracting the DNA, geneticists can zero in on a small region on either the long or the short arm of one of the human genome's twenty-three chromosome pairs.

While the Contursi mutation could have been on any one of the twenty-two non-sex-linked chromosomes (autosomes), it would turn out to lie on chromosome 4. By sheer good fortune, Polymeropoulos was highly familiar with chromosome 4: he had recently linked two other genetic disorders to it—Wolfram syndrome (a rare disorder involving deafness, blindness, and diabetes) and a kind of dwarfism called Ellis–van Creveld syndrome. This work had generated lots of biochemical markers along the chromosome, which guided the pair as they worked. So, within just nine days, they found the locus of the Contursi mutation— chasing it to a small region (band 21) of the long arm (q) of chromosome 4 (all chromosomes have a short arm, p, and a long arm, q). The genetic "zip code" of the Contursi mutation, they concluded, was 4q21.

It took another nine months of painstaking work before they located the precise address within the zip code and sequenced what they thought was the actual mutated gene. Then, says Nussbaum, they got a very lucky break. They checked their sequence against GenBank, a giant open-access computerized database of

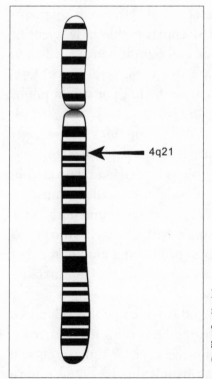

4q21

Figure 13: Nussbaum and Polymeropoulos mapped the Contursi mutation to band 21, on the long arm, q, of chromosome 4. The genetic zip code, or gene locus, is 4q21.

(Copyright © Marie Rossettie, CMI)

gene and protein sequences run by the NIH. Over time, scientists have contributed to this repository some 162 million sequences for over 100,000 distinct organisms. And they got a hit: the mutated gene was a known entity, a gene called SNCA, which coded for a protein called alpha-synuclein. After a lot of hard work and determination, it looked as if they had found a possible key to the mystery of why affected members of the Contursi kindred developed Parkinson's disease.

Alpha-synuclein was sitting in GenBank thanks to the work of two scientists—Stanford's Richard Scheller (who'd found a version of this protein in the nervous system of the Pacific electric ray *Torpedo californica* in 1988) and the University of California, San Diego's Tsunao Saitoh (who reported isolating a fragment of alpha-synuclein in the plaques of Alzheimer's patients in 1993). Like thousands of other scientists in this early scientific crowd-sourcing initiative, Scheller and Saitoh had placed their sequences in the GenBank database as a public service. As Nussbaum explains, "If Saitoh and Scheller hadn't put it in a public database, we would have never found it."

According to Nussbaum and Polymeropoulos, the genetic story behind the Contursi kindred went roughly as follows. SNCA is a gene, and its normal role is to make a relatively obscure brain protein called alpha-synuclein. It's called synuclein, incidentally, to indicate that this protein can be found both in the synapses and in the nucleus. A single base change in the gene's million-letter genetic code, however, produced a mutant form of the protein, which caused affected members of the Contursi kindred to later contract Parkinson's.

On May 27, 1997, Nussbaum and Polymeropoulos submitted a paper to the journal *Science*, listing Duvoisin's team as coauthors, which linked a small mutation in a gene for alpha-synuclein with an aggressive form of Parkinson's disease. One month later—lightning fast for medical research articles—it appeared in print.

I don't remember noticing the *Science* paper at the time. This was still more than a decade before I experienced any symptoms and nearly fifteen years before my diagnosis. But in retrospect, I now realize that what Golbe, Duvoisin, Nussbaum, Polymeropoulos, and their colleagues had found was transformative. No one really expected the Contursi mutation to show up in the DNA of regular Parkinson's patients. And indeed, when neurologists tested their patients, it didn't. This was after all a very rare genetic form of the disease that just affected one family. Nonetheless, this genetic clue offered scientists a wonderful new way of thinking about the disease. The timeline for this form of Parkinson's started long before the affected individuals experienced symptoms or received a diagnosis. The genetic mutation was present at birth and led to a complex biological cascade that years later resulted in an aggressive form of Parkinson's. Unpacking this process would almost certainly hold valuable insights for the wider war on the disease.

And so it did. It happened that Maria Grazia Spillantini, an Italian Alzheimer's researcher working in Cambridge, England, knew of Saitoh's work and had subsequently developed special staining techniques to visualize alpha-synuclein in brain tissue. On a hunch, Spillantini decided to use the stain to search for alpha-synuclein in brain specimens of deceased patients with regular Parkinson's disease. And somewhat surprisingly, even though these individuals lacked the Contursi mutation, she found alpha-synuclein—lots of it. She found it in Lewy bodies.

As we've seen, Lewy bodies, named after the German pathologist Frederick Lewy, are found inside surviving dopamine neurons of Parkinson's sufferers. Today, pathologists everywhere interpret these round masses as confirmation that a patient really had Parkinson's disease. That was how Duvoisin had confirmed the fire chief really had Parkinson's. In patients who have died with Thomas Graboys's aggressive form of the disease, the Lewy bodies were found postmortem not just in the midbrain, which controls movement, but in the cerebral cortex, which handles cognition. But remarkably, despite their pathological importance, in 1997 no one was sure precisely what Lewy bodies were made of.

Spillantini had found the answer: mostly alpha-synuclein. Researchers everywhere took note of this, realizing it might be extremely important. Even though the Contursi mutation doesn't account for the vast majority of Parkinson's cases, the fact that Lewy bodies, the mark of sick and dying dopamine neurons, were stuffed with alpha-synuclein implied that this protein might be a critical player in Parkinson's disease. It suggested that there might be different starting points for Parkinson's that converged in a common biochemical pathway. Unfortunately, Saitoh was not around to read about Spillantini's discovery; he was murdered under mysterious circumstances in May 1996.* But in Germany, the legendary neuroanatomist Heiko Braak, working at Goethe University in Frankfurt, certainly noticed Spillantini's paper in *Nature*. In fact, it was just what he had been waiting for. Braak had started out as a comparative anatomist, dissecting many different animal species, before switching to human pathology and working on Alzheimer's and Parkinson's. Inspired by the discovery that Lewy bodies were made largely of alpha-synuclein, he embarked on a massive Parkinson's disease project, examining the accumulated damage in patients who survived for different lengths of time. Working with autopsy tissue from 168 deceased individuals—some who'd had Parkinson's, some who hadn't—he searched for signs of the disease. Using alpha-synuclein immunostaining, Braak did full-body autopsies of 41 cases of Parkinson's disease, 69 cases with no Parkinson's (but with incidental Lewy

---

*It's a horrifying story. Saitoh arrived home at his La Jolla residence at 11:00 p.m. and was ambushed in what appears to have been an organized hit. He was shot through the window of his car, and his daughter, Loullie, was gunned down while trying to flee. Nothing was stolen. Police have failed to solve the case.

bodies and Lewy neurites), and 58 age-related controls (with no Parkinson's and no Lewy bodies or Lewy neurites). He looked for Lewy bodies and Lewy neurites (deposits in the long axons that project to other nerve cells), and he hunted not only in the brain but in the rest of the body as well. Using Spillantini's powerful new alpha-synuclein stain and a novel technique of examining under the microscope sections of neural tissue that were ten times thicker than those most neuropathologists used, Braak saw clearly what others throughout history had only suggested—that the distribution of Lewy bodies and Lewy neurites was not confined to a few areas of the midbrain. He also discerned something much more profound: that the location of Lewy pathology appeared to change as the disease progressed. Mildly affected cases (people who had died with early-stage Parkinson's disease) showed Lewy pathology in the olfactory bulb of the nose and in part of the vagus nerve—a long projection that connects the gut to the brain. In more advanced cases, he found Lewy bodies and Lewy neurites in the brain stem as well. Still more advanced cases had them in the substantia nigra—marking damage to dopamine cells. The most advanced cases of all displayed Lewy pathology in the forebrain and the neocortex.

Braak argued this was compelling evidence that Parkinson's disease started perhaps decades before any tremor or rigidity appeared—as clinicians and patients had previously noticed. Braak suggested that the disease was possibly triggered by an infection in the gut and/or nose and spread insidiously throughout the brain in six anatomical stages—stages, he suggested, that mapped onto the pattern of symptoms found in epidemiological studies like the Honolulu-Asia Aging Study. Loss of smell and constipation might come in Braak stage 1. REM sleep behavior disorder in Braak stage 2. Classic Parkinson's disease—tremor, rigidity, slowness of movement—showed up in Braak stage 3. Loss of balance in Braak stage 4. And then in Braak stages 5 and 6, the pathology spreads to the forebrain and the neocortex, causing

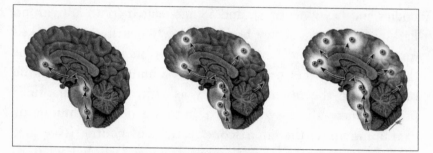

Figure 14: Braak stages 1 and 2—autonomic and olfactory symptoms; Braak stages 3 and 4—sleep issues and motor symptoms (parkinsonism); Braak stages 5 and 6—cognitive and emotional disability (Copyright © Marie Rossettie, CMI)

dementia. The British neuroscientist Chris Hawkes, who has worked with Braak, sums up one of the profound implications of this theory. If Braak is right, then "by the time you go to see a neurologist, you're in Braak stage 3 to 4. The pathology is far more advanced than you'd think from looking at the patient. And to put it crudely, the brain is well and truly pickled."

Braak's theory, published in 2003, was initially met with skepticism, as is the case with so many novel findings in science. But the evidence for it and also for the role of alpha-synuclein would grow and grow. That same year, a group of Mayo Clinic and NIH geneticists announced a landmark discovery in *another* family kindred with an inherited form of Parkinson's disease, a discovery that deepened the connection with alpha-synuclein. Because this extended family had originally settled in Iowa in the nineteenth century, members were known as "the Iowa kindred," although descendants had spread all across America. Over nearly a century, branches of the family had been studied by a series of Mayo Clinic physicians. The geneticist Katrina Gwinn had met one of these clinicians, Manfred Muenter, in the mid-1990s while doing a fellowship at Mayo Clinic's Arizona campus. She'd become fascinated with the kindred and had gotten to know some of them. Realizing the potential importance of this large family for Parkinson's research, she'd invited two British geneticists, John Hardy and Matt Farrer, then working at Mayo's Florida campus, to collaborate. Hardy was a highly accomplished scientist, known in Alzheimer's research for proposing the amyloid cascade hypothesis, which sketched out a sequence of cellular events that led to Alzheimer's disease. Farrer was his postdoctoral researcher. The plan was for them to use the latest gene-hunting technology to discover the secrets of the Iowa kindred—as Nussbaum and Polymeropoulos had done with the Contursi kindred—finding first the locus, then the mutant gene.

The team's first attempt to find the locus of the genetic aberration that caused Parkinson's disease suggested that it lay on the short arm, p, of chromosome 4 in band 15. The team then started the long job of locating what they assumed was a mutant gene somewhere in this locus (4p15). After a couple of years' searching, the team got some stunning news. An unaffected family member, who they had been convinced did not have the genetic mutation, had gone on to develop Parkinson's disease. As he didn't have the 4p15 locus, this called into question everything they had done. They began to suspect that because of a sample mix-up, they had been hunting in the wrong place. Gwinn makes no excuses: "We were wrong—dead wrong."

By now, they had read Nussbaum and Polymeropoulos's 1997 *Science* paper announcing the discovery of the alpha-synuclein mutation in the Italian Contursi kindred. Like Parkinson's researchers everywhere, they checked to see if their patients carried the same mutation. They didn't. They were back where they started.

So they decided to start over. Because genetic linkage analysis depends on having plenty of DNA samples, Matt Farrer and Katrina Gwinn headed out into the field and asked kindred members for more blood. By 2001, the team had enough blood samples to redo the linkage analysis. A new team member, Andrew Singleton, a freshly minted geneticist who'd recently gotten his PhD at Newcastle University in England, took the lead. Singleton's linkage analysis confirmed that Gwinn, Hardy, and Farrer had made a mistake with their first locus and implicated not the short arm, p, but the long arm, q, of chromosome 4. As he tracked the markers, he realized the new locus appeared to include the alpha-synuclein gene found in the Contursi kindred. This was puzzling: they had tested for all known forms of the Contursi mutation and found nothing.

As Singleton pressed harder and harder to find the locus, he noticed some very odd signals. As he put it, "There was something very peculiar about two of the markers that we ran that were in the alpha-synuclein gene region . . . I couldn't figure out what the problem was." Then Singleton got an idea. "It suddenly occurred to me that what could be causing the disease is extra copies of one gene . . . At that point I started to get really, really excited." Singleton demonstrated that unlike the Contursi kindred, this family's Parkinson's disease wasn't caused by a point mutation, an error in the DNA sequence itself. He found that affected members of the Iowa kindred had what geneticists call copy number variation: instead of having only one normal alpha-synuclein gene on each version of chromosome 4, one of the chromosomes had extra copies.

"In Down's syndrome, a child is born with three copies of chromosome 21, rather than two. The concept is really the same here," says Katrina Gwinn, "it's just on a much smaller scale." Singleton's analysis confirmed that family members with Parkinson's had three copies of the normal alpha-synuclein gene—a triplication—on one copy of chromosome 4. On the other copy of chromosome 4, they had the usual single alpha-synuclein gene. Because they had a total of four copies of the gene, instead of the usual two, that meant *twice* as much alpha-synuclein protein was being pumped into affected individuals' bodies. Scientists realized just how significant this was: there was a direct link between quantity of alpha-synuclein and disease. This discovery showed that you didn't need a mutation to get Parkinson's. As the Rush University neuroscientist Jeff Kordower remarked, it showed "the only defect you have to have is too much synuclein and you get Parkinson's."

Hardy describes the news as "a beautiful surprise . . . extremely unexpected. But once you get the result, it makes you understand everything." Other researchers in Europe reported family pedigrees where affected members had both duplications and triplications. The people with the triplications had an earlier onset and much more aggressive illness than those with the duplications. This was also significant. Alpha-synuclein's toxicity depended on dose. The more alpha-synuclein, the worse the Parkinson's disease.

When I think about these breakthroughs now, it's strange to imagine that at one time neuroscientists dismissed the role of genetics in Parkinson's disease. Since the 1997 discovery of the alpha-synuclein mutation, some eighteen potential genetic forms of Parkinson's have turned up. Geneticists are confident that six of them are classically inherited either dominantly or recessively. Dominant mutations include the alpha-synuclein mutations (there are several) and the leucine-rich repeat kinase 2 (LRRK2) mutation—also called dardarin, from the Basque word *dardara*, which means "trembling." The recessive mutations, which are known by the names PARKIN, PINK, and DJ-1, typically have an early age of onset. Most of these are extremely rare.

Researchers are especially interested in LRRK2. It is the only genetic form that is clinically nearly indistinguishable from typical Parkinson's disease. And one variant of LRRK2 (called G2019S) is relatively common: it accounts for about 1.5 to 2 percent of Parkinson's cases in North American Caucasians, about 20 percent of all Ashkenazi Jewish Parkinson's patients, and around 40 percent of cases with North African Berber Arab ancestry.

Most people with Parkinson's like me, for example, don't have a known mutation. But buried in our genomes, there may be sequences that predispose us in some way to develop the disease. Using special gene chips, geneticists have screened the DNA of people with Parkinson's disease and compared it with healthy controls. These gene-wide association studies have found not only a strong correlation with variations in the alpha-synuclein gene but also associations with a mutation that can cause a disease called Gaucher's disease and associations with variations in the gene encoding the tau protein (which is involved in Alzheimer's pathology). Understanding those sequences may . . . well, there's no telling where that knowledge may lead.

The discovery of the association between mutations in the alpha-synuclein gene and Parkinson's had provided clues to how this disease spread throughout the body and brain, launching an explosion of basic research activity. This molecule gave researchers not only a new means of elucidating the complex biological pathways involved in the genesis of the disease but also a way to identify potential targets where a drug might interrupt the process.

The discoveries attracted the attention of the Cambridge scientist Professor Chris Dobson, an expert in protein chemistry. He had noticed a rather similar phenomenon with long-term dialysis patients. In dialysis, a machine filters a patient's blood, trying to do the job of a kidney. It turns out that a serum protein called beta-2 microglobulin is not effectively removed by the dialysis filter. So over a long period of time, the concentration of that protein increases in the patient's blood. At a certain concentration, the dissolved beta-2 microglobulin protein converts into a white sticky deposit (called an amyloid) that accumulates in the patient's bones, joints, and tendons. Just as too much alpha-synuclein is associated with Parkinson's disease, too much beta-2 microglobulin leads to a painful condition called dialysis-related amyloidosis.

Four decades of research had convinced Dobson that proteins were implicated in a range of human diseases—from inherited diseases like cystic fibrosis to neurodegenerative conditions like Parkinson's and Alzheimer's. In a 2004 article titled "In the Footsteps of Alchemists," he had speculated that because many diseases appeared to be connected with misbehaving proteins, one day "it might be possible to block a number of these 'amyloid' diseases with a single drug." If that were true, he added, we might be nearer to realizing "the dream of the medieval alchemists to produce an elixir of life."

WHEN GOOD
PROTEINS GO BAD

Cambridge University has seen more than its share of scientific geniuses. It's the place where Isaac Newton invented calculus and constructed his theory of universal gravitation, where Ernest Rutherford and James Chadwick revealed the structure of the atomic nucleus, and where Paul Dirac enunciated the principles of quantum mechanics. Here, the father-and-son team of William Henry Bragg and William Lawrence Bragg jointly developed the field of crystallography, which facilitated the investigation of living molecules. It's where Fred Sanger sequenced the first protein—insulin—and where Watson and Crick decoded the structure of DNA. To date, Cambridge alumni have won eighty-nine Nobel Prizes.

I'm here to meet another scientific giant—Chris Dobson, master of St. John's College, head of Cambridge's chemistry department, and one of the world's leading experts on proteins. His speculation about an elixir of life has inflamed my curiosity. It's a magnificent day, clear and dry. The May sun lights the historic town like a master cinematographer. Tourists stroll along King's Parade. I'm excited but nervous. I worry that Dobson—who's held faculty positions at Oxford, Harvard, MIT, and Cambridge—will find my questions simplistic.

As I walk to St. John's College, where we're scheduled to

meet, I mentally review what I know about protein chemistry. The headline is that proteins run the show. Life's versatile worker-molecules do just about everything. Proteins build tissues, ferry vital elements around the body, control muscle contractions, and play a central role in the endocrine system and the immune system. And life would be impossible without a special class of regulatory proteins, called enzymes, that catalyze metabolic reactions—speeding them up millions of times.

As high school students learn, the body manufactures all these proteins—some 100,000 varieties—by chaining together smaller molecules called amino acids. Insulin, one of the smallest proteins, is built from 51 amino acids. Titin, one of the largest, contains 33,423 amino acids.

Once the amino acids are chained together, something awesome happens: these long molecules automatically fold into complex three-dimensional shapes—chemists call the shapes "conformations." Imagine a long string of beads crunching up into a series of lumps connected by loops. Folding into this "native" shape is really important. How a protein folds determines what a protein does. Only after the insulin protein, for example, has coiled up correctly can it bind to other molecules and instruct the body to store sugar.

Many gifted scientists have attempted to explain how proteins fold. In the 1950s, the Nobel laureate Linus Pauling figured out that for most proteins there are two preferred basic shapes, or native conformations: a so-called alpha helix, where the protein folds into a right-handed spiral coil; and a (less common) beta sheet configuration, where parallel strands of amino acids aggregate to form the molecular equivalent of a stack of folded cardboard panels. But why and how do proteins select these preferred shapes out of all possible configurations? A protein 100 amino acids long can theoretically fold in a staggering number of ways—10^{143} unique configurations, in fact. If a protein randomly tried all 10^{143} possible conformations, it would take longer than the age of the universe to stumble on its correct native structure. Clearly, nature had found a better way. The NIH scientist Christian Anfinsen argued that the folding "knowledge" was encoded in the order in which the amino acids are strung together—the sequence. This sequence, he reasoned, guided the molecule to adopt the lowest energy state of maximum stability.

In 1961, Anfinsen used chemicals to unfold proteins from their natural state and noticed something remarkable: the molecules spontaneously refolded themselves in a few thousandths of a second. This seemed to prove beyond doubt Anfinsen's theory that the amino acid sequence uniquely specified the structure and therefore the function of a protein. Anfinsen won the 1972 Nobel Prize for his contributions.

But there was also evidence that protein folding wasn't always reversible. Think of what happens when you boil an egg, for example: the albumin protein in the white of the egg unfolds and forms a solid lump. Even after the temperature has fallen, this form persists; the albumin proteins do not refold to the native shape.

When I arrive at St. John's, I'm told to wait in a library in the Master's Lodge. A few minutes later, the master enters. Dressed in a charcoal suit, pin-striped shirt, and blue-patterned tie, the sixty-three-year-old Dobson looks very classy. But he's not at all intimidating. To the contrary, he is friendly and modest. He's even brought two of his younger science protégés, Professor Michele Vendruscolo from Italy and Dr. Tuomas Knowles from Switzerland, to the meeting.

How does a chemist end up working on neurobiological diseases? Forty years ago, Dobson tells me, he began by following in the footsteps of Pauling and Anfinsen. He wanted to solve the many outstanding questions about how proteins fold and misfold. One of the first proteins he analyzed was a common enzyme called lysozyme, which is found in human tears. "It was," he says, "largely a pure scientific quest driven by curiosity." In time, Dobson found that Pauling and Anfinsen's picture of proteins was idealized. It turns out that proteins don't always fold up correctly into their native state. Moreover, many protein molecules keep certain regions unfolded so they can interact with other proteins; this allows one protein to have several functions.

But more significantly, Dobson's research program became driven less by pure curiosity and more by human need. "In the mid-1990s, a London physician, Mark Pepys, visited me in Cambridge," Dobson explained. "Pepys was convinced that defective lysozyme proteins were killing some of his patients." Those unfortunate individuals had inherited a rare mutation that caused them to produce abnormally folded lysozyme molecules. Instead of remaining in their native alpha helical form, these proteins went rogue and unfolded, exposing the molecule's sticky backbone. Then they joined with other lysozyme proteins to form long thin fibrils that aggregated into tough white clumps—or amyloids—kilograms of which accumulated in diverse organs from the liver to the brain. Patients became sick with an incurable condition called hereditary systemic amyloidosis.

As far as Dobson could make out, "the mutation made the proteins less stable . . . so they had a greater than average tendency to unfold. And when the mutant proteins unfolded, they didn't fold back again as Anfinsen had found; they stuck to other lysozyme proteins, causing havoc." The encounter changed the direction of Dobson's work. The origin of the patients' lethal pathology, he realized, was basically a matter of chemistry—as in the boiled egg, a protein had shifted from one shape, where it was dissolved in water, to another, where it became insoluble. Dobson was determined to understand why.

Using chemical agents and heat energy, he stressed a number of common proteins, including lysozyme, in a test tube and found it was relatively easy to unfold the molecules. And once unfolded, he discovered, the proteins were remarkably vulnerable (given enough time) to converting to the amyloid form. The process involved misfolded molecules morphing into long thin "fibrils" that stick together and grow into clumps. Says Dobson, "The amyloid state is highly organized and very stable; it's perfectly normal chemistry. It was not invented by biology; it was not invented by these diseases. The real question is not why it had happened in a few of Dr. Pepys's patients, or in my test tubes, but why such amyloids almost never build up in healthy living cells."

Dobson gets excited as he answers his own question. "Nature has been extremely cunning. She hasn't left proteins to fend for themselves." Proteins, he explains, reside inside cells—complex structures jam-packed with ingenious control systems to prevent molecules from misfolding. For example, cells come equipped with armies of special helper molecules called "chaperones" whose function is to assist newly synthesized proteins to fold correctly. One chaperone molecule, known as GroEL, even contains a protected inner chamber where a partially folded protein molecule can hang out while making final conformational adjustments. If for any reason proteins still misfold, or fail to refold, the cell is equipped with special garbage disposal machines called proteasomes and lysosomes that quickly remove any misfolded protein, chop it up, and recycle the parts before damage is done to the host organism.

Despite all these cunning mechanisms, says Dobson, for multiple reasons (genes, environment, age, and more) proteins still misfold, and over time this leads to amyloid diseases, which are remarkably widespread. Even though each disease involves a different protein—alpha-synuclein is involved with Parkinson's, tau and amyloid-beta with Alzheimer's, huntingtin with Huntington's disease, superoxide dismutase with some forms of amyotrophic lateral sclerosis—the cellular control systems fail in much the same way.

But there was a missing piece to this theory. It was one thing for a protein to misfold and form aggregates in a single cell. But how exactly can rogue proteins proliferate like viruses or bacteria? How can they self-replicate and spread from cell to cell like an infectious pathogen?

The answers to these questions lead us back to a fascinating biological heresy, a tale featuring a controversial scientist named Stanley Prusiner. Prusiner's remarkable scientific odyssey—which at first sight has little to do with Parkinson's—started in 1972, shortly after he'd qualified as a doctor. He watched one of his patients die of a rare and invariably fatal condition called Creutzfeldt-Jakob disease and decided to investigate the malady. In this rapidly progressing disease, patients suffer dementia, memory loss, hallucinations, and more. Prusiner quickly discovered that fellow scientists had linked Creutzfeldt-Jakob disease to two other infectious neurodegenerative diseases: scrapie, a disease known since the 1720s that affects sheep and goats with a kind of animal dementia; and kuru, a disease of the Fore tribe in New Guinea, where members had a cannibalistic ritual of eating the brains of deceased tribal members. Prusiner noted that the three mammalian diseases had much in common. All were 100 percent fatal. All left sponge-like holes in their victims' brains (leading to the label "spongiform"). All killed without evoking an immune response. All required long incubation times—generally measured in years. All appeared to be contagious: when brain tissue from deceased sheep or people was injected into healthy animals, the recipients got sick.

But most remarkable was that this set of diseases appeared to be carried by a pathogen unlike anything seen in the history of medicine. The mysterious entity was very, very difficult to kill. Scrapie brain tissue, for example, remained infectious even after being frozen, boiled in water, soaked in formaldehyde, exposed to ionizing radiation, and flooded with intense ultraviolet light—processes that were known to rapidly destroy the DNA and RNA inside pathogens like viruses and bacteria. There were reports of contaminated fields being cleared of sheep and abandoned for ten years; remarkably, when new flocks of sheep were allowed to graze there, the previously healthy animals came down with scrapie.

In the 1980s and 1990s, scientists found four other diseases that behaved like scrapie, kuru, and Creutzfeldt-Jakob disease: bovine spongiform encephalopathy (BSE), or mad cow disease; a new variant of Creutzfeldt-Jakob disease resulting from eating BSE-diseased cattle (vCJD), something that had caused a massive public health scare in Britain; and two very rare hereditary diseases, fatal familial insomnia and Gerstmann-Sträussler-Scheinker disease.

Thinking this group of seven diseases must be caused by a very slow-acting and hardy virus, Prusiner spent years trying to isolate the infectious agent. Try as he might, he found no virus. So in an act of scientific heresy, he reasoned the pathogen was not a traditional infection, caused by a bacterium, virus, or fungus. The disease, Prusiner claimed, was directly spread by proteins—not just any proteins, but infectious ones, which he labeled with a catchy new term: prions. Many scientists were skeptical at the idea of this contagious chemistry. As Prusiner put it, "It is difficult to convey the level of animosity that both the word 'prion' and the prion concept engendered. At every turn, I met people who were genuinely irritated by my findings." To be sure, it was a disruptive idea. To spread disease, prions would have to be able to self-replicate inside the mammalian body—something that biological dogma held was exclusively the province of DNA and RNA. Prusiner was implying that prions worked a bit like in a zombie movie, where afflicted zombies turn healthy people into zombies just by touching them. In the mammalian body, Prusiner hypothesized, the misfolded prion protein replicated by "coercing" normal proteins in the vicinity to adopt its shape, and those proteins, in turn, converted other normal proteins in a chain reaction.

But was it true? The same year that Bob Nussbaum and Mihael Polymeropoulos sequenced alpha-synuclein from the blood of members of the Contursi kindred, the University of California, San Francisco, neuroscientist Stanley Prusiner received the 1997 Nobel Prize for the "discovery" of prions. It was one of the most controversial decisions ever made by the Nobel committee, given that many scientists considered the notion of prions heretical. Dr. Lars Edstrom, a member of the awards committee, admitted, "There are still people who don't believe that a protein can cause these diseases, but we believe it."

Dobson and other chemists speculated about the link between rare prion diseases and more common conditions like Parkinson's and Alzheimer's. Perhaps the same basic shape-shifting tricks that prions display can be used to explain all amyloid diseases, although in Alzheimer's, Parkinson's, and amyotrophic lateral sclerosis the disease spreads only from cell to cell *within* the animal. By contrast, in prion diseases like scrapie the misfolded protein is contagious and can jump from individual to individual and even cross species barriers—in the case of mad cow disease from sheep (scrapie) to cow (BSE) to human (vCJD).

So what does it mean to say that neurodegenerative diseases can spread from neuron to neuron in this way? What does it explain about the timeline of the malady? I pose this question to Dobson and the younger researchers, who join in the conversation. Tuomas Knowles, who studied physics in Zurich and Geneva before coming to Cambridge, suggests a chemical metaphor: "Think of it like crystallization. I make a solution of high-concentration copper sulfate, hang a string in it, and leave it and wait. And nothing seems to happen for days. And then suddenly, when I'm about to give up, crystals start to form—ones I can actually see." I remember trying to grow crystals as a boy, and he's right: it took much longer than I expected. By analogy, Tuomas explains, years before patients suffer any symptoms, proteins somewhere in the body were unfolding and forming an amyloid nucleus.

Michele Vendruscolo, a graduate of the University of Trieste, extends the metaphor: "The initial stage of crystallization is called nucleation, and that may take time. But once it's going, there are shortcuts. If you break off a bit of Tuomas's crystal and drop it in my crystal-growing solution, it acts like a seed or template, and it makes the process go faster. In this way, after a long incubation period, a purely physical entity can spread like an infection."

According to this trio of brilliant physical scientists, something similar seems to be happening with all amyloid diseases: misfolded single proteins (monomers) stick to other molecules to form oligomers, which grow into fibrils, which become amyloid plaques. Along the way, growing fibril structures can break off and serve as templates for secondary amyloid growth. The secondary spread of fibrils is quicker in pure prion diseases like scrapie; that's what may account for prion diseases' animal-to-animal contagiousness. But the idea is the same for noncontagious diseases like Parkinson's. And compelling evidence that alpha-synuclein could spread in a prion-like manner in fact emerged in 2007, data that persuaded neuroscientists as well as chemists.

That year, the Swedish scientist Patrik Brundin and his colleagues examined the brain of one of the Lund neural grafting patients, who had died in 2005. The patient's grafted fetal tissue had been implanted in two different operations, eleven and sixteen years earlier. What Brundin and his colleagues observed was astonishing. Some of the grafted fetal neurons now stained positive for Lewy bodies. The neurons originally derived from a fetus a few weeks old had developed the pathological signs of an old person's disease. During the coffee break at a conference at the Karolinska Institute that summer, Brundin showed pictures of the brain tissue slides on his laptop to the neuropathologist Jeff Kordower. Kordower was stunned. He had previously examined the brain of a patient who had died just four years after fetal graft surgery that had no such pathology, so, he surmised, perhaps the spread of misfolded alpha-synuclein needed longer to manifest as Lewy bodies. Kordower realized he had a way to test this. Back in Chicago, he had the brain of a sixty-one-year-old female patient who had died fourteen years after graft surgery. So Kordower returned to his laboratory at Rush University Medical Center and stained the tissue; sure enough, he found that the naive fetal neurons after fourteen years contained Lewy bodies.

Brundin then called the neuropathologist Tamas Revesz in London, who he knew had the brain of yet another patient who had received transplants in Lund, Sweden, more than a decade before dying. When Revesz also found Lewy body pathology in the brain of this third neural grafting case, Parkinson's researchers became very excited. Taken together, these findings suggested two compelling conclusions. First, they indicated that the fetal transplants did not stop the progression of the disease; even *after* the transplanting of new cells, the disease process continued. Second, they suggested that alpha-synuclein was truly capable of jumping from cell to cell in a prion-like fashion. Given enough time, the misfolded protein could spread throughout the brain. The results were published in two back-to-back articles in *Nature Medicine*. This was somewhat of a paradigm shift, and a new era in Parkinson's research started.

An obvious question: Why has evolution "allowed" protein misfolding to happen at all? After all, cells possess an elaborate series of control mechanisms to help proteins fold correctly and to recycle them if they don't. These cellular defenses are not perfect. They can fail, Dobson says, for many reasons, including genetic mutations and toxic damage to the cell (from pesticides or air pollution, for example). But the biggest factor, he says, is time. "One remarkable thing about amyloid diseases like Parkinson's and Alzheimer's is that they very rarely are diagnosed in people under the age of sixty." Evolution, he argues, cares that we live long enough to pass on our genes to our offspring, but it doesn't really care after that. "And so it's evolved proteins that are stable enough and protected well enough by cellular defense mechanisms to last forty, fifty, or sixty years, but there's not much margin of safety."

Until very recently, there didn't need to be. Since human beings first walked the earth a few million years ago, life spans were short, rarely exceeding twenty-five years. In the seventeenth century, when Isaac Newton lived and worked in Trinity College, the average life expectancy at birth was just thirty-seven years. While the low number results in part from very high childhood mortality, in that era relatively few people made it to sixty. A child born today in an advanced Western society has a life expectancy of nearly eighty. By 2040, in countries like the United States and the United Kingdom, one in five people will be sixty-five or over. The numbers of cases of neurodegenerative diseases like Parkinson's and Alzheimer's will soar.

But there's good news. Dobson believes these protein-folding diseases will be easier to cure than cancer. "All we've got to do is find a way to help evolution. If you are trying to stop crystallization phenomena in a test tube, you reduce the concentration of the liquid. By analogy, to slow Alzheimer's and Parkinson's, you need to reduce the amount of beta amyloid and alpha-synuclein." While the biology of Parkinson's disease may be fiendishly complicated, the nonbiological origins are relatively simple. Teams are working on small molecule drugs and antibodies. A multidisciplinary team of German researchers tested twenty thousand compounds searching for agents that blocked alpha-synuclein molecules from assembling into toxic aggregates. One substance, named Anle138b after the chemist Andrei Leonov, who synthesized it, has proved remarkably effective in mouse models of Parkinson's disease. It readily crossed the blood-brain barrier, caused no adverse effects at high doses, and significantly reduced oligomer accumulation. As a result, the researchers claim, Anle138b-treated parkinsonian mice experienced less nerve cell degeneration and survived for much longer than untreated controls.

The Contursi kindred, alpha-synuclein, Braak's staging scheme, the Iowa triplication cases, prions and amyloids . . . it was quite a story: a tale that suggested that one day this basic research would yield clinical weapons to vanquish Parkinson's and other neurodegenerative diseases.

I find the notion that the disabling symptoms of Parkinson's disease that I and other Parkies experience are caused by toxic species of alpha-synuclein spreading prion-like throughout the brain to be a very powerful one indeed. In a story with many setbacks, this body of research gives me genuine hope, for it suggests that in theory chemical interventions to break up and destroy the misfolded protein aggregates might help slow, stop, or reverse Parkinson's. If given early enough, such treatment might even prevent the disease from ever reaching clinical significance.

Some neuroscientists are as excited about drugs to dissolve alpha-synuclein as they once were about fetal grafts and growth factor therapy. But this time, scientists are cognizant of the fact that unless they get all the details correct, a potentially breakthrough therapy might fail to prove safe and effective in clinical trials. So before we discuss the next best hope emerging from the lab, let's anticipate the challenges that lie ahead, starting with a visit to Sun City, Arizona.

13

DAMAGE ASSESSMENT

In 1960, the developer Del E. Webb chose Sun City, Arizona, as the location for the first commercial retirement community in the United States. Half a century later, the region has become a prime area for senior citizens who are looking for a comfortable place to spend their golden years. And that makes it a perfect spot to study Parkinson's and Alzheimer's, according to the pathologist Tom Beach, a leading scientist with the Arizona Study of Aging and Neurodegenerative Disease. At any one time, Beach and his colleagues at Banner Sun Health Research Institute are tracking a cohort of some nine hundred to a thousand volunteers from various Sun City retirement communities. The institute provides volunteers with annual specialized neurological and physical exams. For their part, the retirees donate their bodies for future scientific research after they die.

Nowhere in the world are autopsies carried out so speedily as in Sun City. "We have on-call autopsy teams ready to act 24/7," says Beach, and "their goal is to perform the autopsy within a three-hour window after death." Time is critical, because within minutes of death, vital biochemical evidence (from RNA to gene expression) starts to degrade. If you want to accurately assess the degree of damage in the fresh brain tissue in patients dying with neurodegenerative disease, then Banner is the go-to place.

The bodies of the Sun City residents who signed on to participate in this program are brought to the institute's morgue, where two teams of technicians work to collect specimens—one team harvesting the brain, the other the rest of the body. Technicians on the brain team remove the brain from the skull and dissect out the brain stem and cerebellum. Then, using a long instrument resembling a bread knife, they slice the cerebral hemispheres into more than a dozen coronal sections. Half of the sections are frozen in dry ice and stored in freezers so that they will be available to researchers all over the world.

Samples from the other half of the brain, and from more than forty distinct body parts (including the heart, lung, liver, kidneys, and eyes) dissected by the other team, are preserved in formalin and stored in carefully labeled plastic Tupperware containers for anatomical study. When pathologists put in a request to Banner, technicians retrieve the appropriate tissue, embed it in wax, and mount it on a special machine called a microtome—a device that shaves gossamer-thin wax slices of the specimen. Banner technicians drop the shavings in water and scoop them up onto glass slides, slides that scientists can stain with antibodies for chemicals like alpha-synuclein so the anatomical and pathological details can be visualized under light microscopes.

To date, the institute has performed 1,600 rapid autopsies, of which about 220 are from diagnosed Parkinson's patients, and this activity has generated an archive of hundreds of thousands of tissue samples. A skilled neuropathologist can deduce a lot from such tissues. Perhaps most remarkably, Beach sees an alarming number of diagnostic mistakes—cases displaying Parkinson's-like symptoms that do not in fact have regular Parkinson's disease. In a recent study comparing clinical diagnosis with postmortem tissue analysis, the Mayo Clinic neurologist Charles Adler, Tom Beach, and their colleagues reported sobering results. Clinicians were 88 percent accurate diagnosing Parkinson's disease in patients who had had symptoms for more than five years, but their accuracy rate dropped to only 53 percent for patients who had signs or symptoms for less than five years. How are they so sure? Because even though these Sun City retirees in life displayed the clinical syndrome of parkinsonism, the meticulous postmortem analysis revealed that about half of them lacked the expected pathology involving Lewy bodies in the dopaminergic areas of their brains. And if there are no Lewy bodies, says Beach, "it's not Parkinson's disease."

This is a somewhat puzzling issue to get your head around. There's parkinsonism the syndrome, and then there are the underlying diseases that cause the syndrome—which include not only regular "sporadic" Parkinson's disease (the most common kind) but also other Parkinson's-like conditions (resulting, for example, from a rare genetic mutation or a toxin like MPTP), which to a pathologist are simply different diseases. Recall, for a syndrome to rise to the level of a disease, you need to know either how the condition starts (the cause) or how it ends. No one knows for sure the cause of regular Parkinson's disease, but we do know how it ends. Because patients with the most common form of Parkinson's invariably end up with Lewy bodies—which are made up largely of misfolded alpha-synuclein—pathologists like Beach refer to it as a synucleinopathy.

The fact that some Sun City retirees with parkinsonian symptoms turn out to have little or no alpha-synuclein in their brains cannot be ignored. It implies that they had a different underlying disease. In some of these cases, the retirees' brains turn out to contain not Lewy bodies but a different misfolded protein called tau, which is also found in Alzheimer's disease. According to Beach, aggregates of misfolded tau protein (or tau tangles, as they are called) are found in an aggressive Parkinson's-like disease called progressive supranuclear palsy (PSP), the disease that killed the comedian Dudley Moore.

Clinically, PSP presents somewhat differently from the regular or sporadic Parkinson's disease that I have. But the differences are subtle if you're not trained to look for them. In PSP, patients often report having problems moving their eyes up and down. Also, sufferers don't generally respond to L-dopa and other do-paminergic drugs. Without such medicines, patients become immobile and at risk for complications like pneumonia, and PSP cases tend to die usually by seven or eight years after diagnosis. But while a clinician might confuse early PSP with Parkinson's, a pathologist has no such confusion. The presence of tau tangles makes it a tauopathy, which to a pathologist means it's by defini-tion driven by a different disease process.

There are several other tauopathies that produce a syndrome of parkinsonism, including encephalitis lethargica—the type of parkinsonism that broke out after the 1918 influenza epidemic, which Oliver Sacks wrote about in *Awakenings*—and dementia pugilistica, the parkinsonism that afflicts individuals who have suffered head trauma, often as a result of high-impact sports, like boxing (with Muhammad Ali) and American football, the latter of which has received extensive media coverage recently.

Then there are still other pathologically distinct forms of Parkinson's-like conditions—for example, multiple system atro-phy (MSA), where the alpha-synuclein aggregates appear in glial cells rather than in neurons. Clinically, MSA doesn't always re-spond to levodopa. It progresses more rapidly than regular Parkin-son's but is less likely to cause dementia.

There is a variety called vascular parkinsonism, which scien-tists believe results from multiple small strokes that affect the basal ganglia. Even more confusingly, some (mostly very rare) genetic forms of very early-onset parkinsonism (including the so-called PARKIN, PINK, and DJ-1 mutations) do not seem to present with Lewy bodies at all. In other words, they may generate a disease that looks similar to Parkinson's but that works via a distinct pathological process. In these individuals, the pathology that destroys the substantia nigra seems to spread via mitochondrial damage (mitochondria, recall, are the cellular power stations).

But the relatively common leucine-rich repeat kinase 2 mutations are even more puzzling. Pathologists have found that individuals with *this* mutation can display at postmortem either alpha-synuclein aggregates (Lewy bodies), tau aggregates, or no aggregates at all. In the clinic, a skillful neurologist can discriminate those LRRK2 carriers with alpha-synuclein inclusions (which present with conventional Parkinson's disease) from those with tau pathology, which develop as PSP cases.

It's not hard to see why, especially in the early stages of disease, clinicians often confuse regular sporadic Parkinson's disease with one or more of these similar-looking but pathologically distinct Parkinson's-like conditions. Indeed, the study from Adler, Beach, and their colleagues that estimated an error rate of around 50 percent for cases with symptoms less than five years is quite alarming and points to this difficulty in diagnosing the right type of Parkinson's. This is higher than previous estimates in Sweden and the United Kingdom, which put the error rate at around 20 percent. Beach argues that even if the rate is lower, diagnostic errors present a significant challenge. Indeed, they may be one reason why recent efforts to demonstrate disease-modifying interventions such as neurotrophic factors have been so unsuccessful. Says Beach, "It's a major roadblock. How can you run a clinical trial if 50 percent of the people don't have the disease you're trying to treat?"

Another complicating factor is that elderly people like those in the Sun City cohort turn out at postmortem to have more than one disease. As Beach explains, "Someone might die with the amyloid beta and tau tangles of Alzheimer's disease, with Lewy bodies from Parkinson's, with infarcts [areas of dead tissue rendered necrotic by an obstruction of local blood supply], and more." It seems that with advancing age the number of diagnoses of pathological conditions goes up exponentially. So, people dying in their eighties, like many of the folks at Sun City, have twice as many diagnoses as people dying in their mid-seventies. As Beach notes, "Most people die with four or five diagnoses, and some people have twelve diagnoses when I examine their brains."

And for those patients who really did have regular sporadic Parkinson's at death, the evidence carried in their tissues goes a long way to explaining the variation in symptoms and speed of progression experienced by different individuals (such as Pamela Quinn and Tom Graboys). Beach's colleague the German neuropathologist Heiko Braak had reported in 2003 that in Parkinson's disease the pathology appears to spread out through the brain in stages, bringing new symptoms in its wake. The Lewy bodies and Lewy neurites, he suggested, started in both the body (for example, in the gut) and the brain (probably the olfactory bulb) before spreading to the brain stem, substantia nigra, and cortex. Beach thinks that the disease starts in the brain before it is found in the body. As he notes, "The olfactory bulb may be where it all begins." From there, Beach says, "it can go the route that Braak says to the medulla, the brain stem, and on to the substantia nigra and to the cortex."

A somewhat demoralizing consequence of the Braak theory, one that Beach and I talked about in Sun City, is that by the time most people are diagnosed with Parkinson's disease they are already in Braak stage 3 or 4 with 70 percent loss of nigral neurons. And this extensive damage may be another reason why the disease-modifying trials have so far had limited success. In a landmark study involving a collaboration between the Arizona Study of Aging and Neurodegenerative Disease and Rush University, the neuroscientists Jeff Kordower, Tom Beach, and their colleagues investigated twenty-eight postmortem parkinsonian brains and compared them with nine healthy controls. Because these Parkinson's patients survived for varying lengths of time after diagnosis (from one to twenty-seven years), the researchers set out to estimate dopamine loss in the substantia nigra and putamen at different stages of the disease.

The researchers searched for signs of damage in the striatum by using stains for two proteins normally active at dopaminergic terminals. The team found plenty.

At diagnosis, the substantia nigra pars compacta and the dorsal putamen had lost more than half of their functional markers. But as the slide graphically shows, the news gets much worse. By four to five years post-diagnosis, the two stains are virtually gone. The lack of such dopamine markers in the striatum—the main input to the basal ganglia—suggests that within five years of getting a diagnosis from a neurologist, a patient has virtually no functioning dopamine nerve endings left. Unless there are nerve endings still present that did not stain for the dopamine markers, these results imply the disease is so advanced by this stage that there may be very few nerve terminals left to save.

This is not the kind of news that people with Parkinson's disease want to hear. Given that I was diagnosed in 2011, my five years will be up in 2016.

This news was sobering for researchers as well. For Jeff Kordower, it provided perhaps the most telling explanation for why his promising gene therapy work with Ceregene to deliver growth factors to the brain (which I detailed in chapter 9) had failed in human trials. Says Kordower, "The data was so striking that I didn't believe it, and I held on to the data for two years, until I was able to get another cohort from Australia to check it against. But the results were the same: by five years, there were no fibers left in the striatum."*

---

*Not everyone agrees that matters are hopeless. The Swedish neuroscientist Anders Björklund published a paper in 2012 in *Science Translational Medicine* in which he showed in an animal model that alpha-synuclein switched off the GDNF receptor-signaling pathway. The pathways could be restored, in theory, with nuclear receptor related 1 protein, or NURR1. This indicates that if you could just put the receptors back online, then GDNF could work.

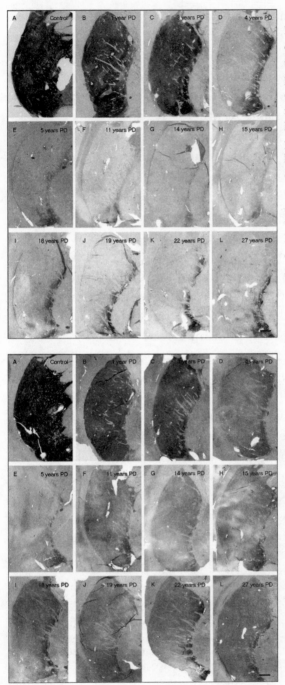

Figure 15: This slide compares the presence of tyrosine hydroxylase (a protein normally active at dopaminergic terminals) in the putamens of a healthy control and patients who died after having had Parkinson's disease for between one and twenty-eight years. After five years, the tyrosine hydroxylase has largely vanished. (Kordower et al., *Brain* (2013) 136 (8): 2419–2431)

Figure 16: This slide compares the presence of dopamine transporter (a protein normally active at dopaminergic terminals) in the putamens of a healthy control and patients who died after having had Parkinson's disease for between one and twenty-eight years. After five years, the dopamine transporter has largely vanished. (Kordower et al., *Brain* (2013) 136 (8): 2419–2431)

He concludes, "I've learned that it's really complicated trying to repair the human brain . . . The challenge is far greater than we anticipated back when we started." If this finding that the dopaminergic damage is really so advanced at five years is confirmed, there are profound consequences for future trials of some potential disease-modifying interventions for Parkinson's. These data imply that even if basic researchers come up with exciting new weapons that could in principle succeed where growth factors and neural grafts have so far failed and combat the toxic process going on in my brain, it may not be enough. Clinical investigators may still have to confront serious problems demonstrating efficacy in controlled clinical trials of most patients with established Parkinson's disease. This fascinating conundrum has driven some researchers to search for novel ways of bringing new therapies to market.

The ultimate goal of biomedical research is to deliver safe and effective therapies to patients. To find new treatments, every year the NIH spends some $30 billion on biomedical research, funding largely academic research. The private sector—including the medical device industry, "big pharma," and the biotech sector—chips in another $70 billion. New drugs cost so much to develop—on average it takes about $1.3 billion and fifteen to eighteen years to bring a drug for a central nervous system disorder like Parkinson's to market—because most new drugs fail. According to the Princeton health economist Uwe Reinhardt, commercial drug development is "like drilling for oil. You hit a lot of dry holes before you hit a gusher. So the cost of all the failures has to be charged, ultimately, to the successful drug."

Over the years, there have been novel efforts to reduce the costs of failure; one of the most intriguing experiments in which Parkinson's researchers have lately made some headway is "drug repurposing." The concept was originally discovered by chance, as so many things are in science. In 1952, clinicians noticed that patients given a tuberculosis drug called iproniazid were "inappropriately happy." In 1958, the drug was reborn, or "repurposed," as one of the first-ever antidepressants. A more famous example involves the pharmaceutical company Pfizer. In the 1980s, Pfizer researchers tested a new medicine intended to treat angina (a chest pain resulting from blockages of the coronary arteries) called UK-92,480 in the Welsh town of Merthyr Tydfil. The drug was a bust. It did little to combat angina. Pfizer prepared to declare failure, to rack up another "dry hole." But then something surprising happened. Men in the trial reported an intriguing side effect of the drug—enhanced erections. Pfizer repurposed UK-92,480 from an angina medication to a blockbuster erectile dysfunction drug called Viagra . . . and the rest is history.

Over the years, a number of other drugs—some failures, others successes—have been famously repurposed. Thalidomide, a tragic failure as a morning sickness drug (tragic, of course, because thousands of babies were born with malformed limbs), now treats leprosy and multiple myeloma. The hypertension drug minoxidil turned out to control hair loss. Methotrexate, developed as a chemotherapy against multiple forms of cancer, has found a new purpose in treating rheumatoid arthritis. And aspirin has an ever-growing list of uses—painkiller, blood thinner, anti-inflammatory, diabetic cataract preventer, even for the prevention of breast cancer. The logic that the same drug can find multiple uses is compelling.

A large number of already approved drugs (more than two thousand) sit on drugmakers' shelves. And many scientists have no doubt that some of them could prove safe and efficacious against Parkinson's disease. Because repurposing can use preclinical safety data that were gathered at the time of the original drug application, it saves time and money and reduces the risk of failure. An estimated 25 percent of repurposed compounds tested in phase 2 trials make it to market (compared with 10 percent of new chemical entities), so it's no wonder that scientists see this area as one worth investigating.

The easy part of repurposing is finding candidate compounds. In 2010, the British charity Cure Parkinson's Trust, led by Tom Isaacs, whom we met in chapter 5, became interested in applying the concept of repurposing systematically to Parkinson's disease. It supported the creation of an international team of researchers led by the neuroscientist Patrik Brundin with a clearly articulated goal: to screen thousands of FDA-approved drugs and over-the-counter supplements for anti-Parkinson's activity and test the medicines in patients via a worldwide clinical trial network. These small "learning trials" would run in parallel, hence the name the Linked Clinical Trials (LCT) initiative.

In the first phase of the LCT project, researchers identified seventy-two candidate compounds used against a wide range of conditions, including type 2 diabetes, hypertension, thalassemia, and cancer. Their top five choices included three type 2 diabetes drugs (exenatide, liraglutide, and lixisenatide), an iron chelator (deferasirox), and a statin (simvastatin). In the first learning ("open label") trial to be completed, Iciar Aviles-Olmos and colleagues reported that exenatide (an injectable type 2 diabetes medication) produced a statistically significant effect in slowing the progression of symptoms measured on the UPDRS and in a rating scale measuring cognitive function. One major weakness of the trial was that there was no placebo given to the control group because of the excessive expense of making injection pens containing a non-active substance. At least to avoid observer bias, the patients were filmed and their symptoms scored by a neurologist in a remote location who was not aware if they were getting exenatide or not. Interestingly, even when the patients had been taken off the experimental drug exenatide for one full year, they remained significantly better than the control group. This suggests that twelve months of exenatide treatment actually promotes some kind of long-term structural change in the parkinsonian brain or even protects neurons from death. Future larger trials will have to address whether exenatide actually modifies Parkinson's disease progression after any placebo effect has been accounted for. But it's a promising start.

Unfortunately, the odds are that many repurposed drugs will ultimately fail. A decade ago, the National Institute of Neurological Disorders and Stroke started a repurposing program called the Committee to Identify Neuroprotective Agents for Parkinson's (CINAPS). CINAPS looked for existing FDA-approved drugs that were potentially neuroprotective for Parkinson's disease. The scientists selected four candidates as having the right properties to be successful (for example, the ability to cross the blood-brain barrier): the supplement coenzyme Q10; an erectile dysfunction treatment, GPI-1485; an antibiotic, minocycline; and creatine, an energy supplement used by bodybuilders. They eliminated three in small trials known as futility studies, which showed these agents did not work. The team decided to go forward with the remaining chemical agent, creatine. It mounted a large phase 3 study testing a purified form of creatine involving 1,720 people with early-stage Parkinson's at fifty-one medical centers in the United States and Canada. In 2013, the researchers reported that the patients taking creatine were doing no better than study subjects receiving placebos.

Even cheaper repurposed drugs like creatine must confront the many methodological challenges of the large phase 3 trial, from the fact that the disease is already advanced at diagnosis to the diagnostic errors clinicians make, from the difficulty of recruiting and retaining a suitable patient cohort to the pervasive influence of placebo effects, from the insensitivity of the UPDRS to the masking effect of L-dopa, and more.

As the deputy director of NINDS, Walter Koroshetz, says, "We just have to do something different to increase our chances of success . . . We can't just keep throwing stuff out with limited scientific rationale into these large expensive trials and failing . . . We just have to do something better."

According to Koroshetz, that comes down to two big ideas. The first relates to the concept of being able to observe and measure the disease in the living brain—an idea that involves finding disease "biomarkers." The Sun City neuropathologists had been able to tell at postmortem precisely what kind of disease a patient died with. Neuroscientists needed effective ways of measuring disease while people were still alive. The second big idea is to test new treatments on patients who have just been diagnosed . . . or, ideally, even earlier. For as we learned in Sun City, by five years after diagnosis the damage to some areas of the brain, like the dopamine system, may be too extensive to rescue.

Fortunately, researchers studying Alzheimer's, the most common neurodegenerative disease, have come up with some powerful and exciting strategies to detect and treat disease at an early stage. And these ideas might just work for testing new Parkinson's therapies as well.

14

LEARNING FROM ALZHEIMER'S

On November 25, 1901, Auguste Deter, the wife of a railway office clerk, was admitted to the Hospital for the Mentally Ill and for Epileptics in Frankfurt, Germany. Deter, who presented with memory loss and delusional behavior, was examined by a thirty-seven-year-old psychiatrist and neuropathologist by the name of Alois Alzheimer. The clinical notes he made that day have been preserved, which has allowed researchers to peer into the past. So we can eavesdrop on the examination of the first case of what is now termed Alzheimer's disease.*

Alzheimer: *What is your name?*
Deter: *Auguste.*
Alzheimer: *Family name?*
Deter: *Auguste.*
Alzheimer: *What is your husband's name?*
[She hesitates, then finally answers.]
Deter: *I believe . . . Auguste.*
Alzheimer: *Your husband?*
Deter: *Oh, so!*

---

*Auguste Deter's medical notes are available in German at www.deutsche-alzheimer.de /unser-service/archiv-alzheimer-info/auguste-d.html.

Alzheimer: *How old are you?*
Deter: *Fifty-one.*
Alzheimer: *Where do you live?*
Deter: *Oh, you have been to our place?*
Alzheimer: *Are you married?*
Deter: *Oh, I am so confused.*
Alzheimer: *Where are you right now?*
Deter: *Here and everywhere, here and now, you must not think badly of me.*

After Deter died in 1906, Alzheimer examined her brain and found strange formations, the amyloid plaques and neurofibrillary, or tau, tangles that we encountered in the last chapter. Just as Lewy bodies came to characterize Parkinson's disease, these plaques and tangles would come to be accepted as the defining pathological hallmark of Alzheimer's disease, a malady Alois Alzheimer dubbed the "Disease of Forgetfulness" at the time.

Today, Alzheimer's disease is the sixth-leading cause of death in the United States, afflicting more than five million Americans and over twenty-six million individuals worldwide. People with Alzheimer's progressively lose memory, language skills, and the ability to accurately perceive time and space. Like Deter, they will eventually require full-time care. The risk of developing Alzheimer's also increases with age. At age sixty-five, on average, one in one hundred people have it. By age seventy-five, on average, one in ten people have been diagnosed. And among ninety-year-olds, on average, one in three develop the condition. Because the world's population is aging fast, researchers predict that Alzheimer's will become much more common in the future.

As with Parkinson's, neuroscientists think that the underlying disease of Alzheimer's starts a decade or more before diagnosis. While the exact mechanism by which the disease spreads isn't known, the pathology is thought to be driven by the aggregation of misfolded proteins—either amyloid-beta plaques (named after the protein amyloid beta from which it is composed) or tau tangles, or both—with the damage spreading from neuron to neuron in a prion-like manner. Researchers also believe that Alzheimer's disease starts in the region of the hippocampus (a brain area that is critical for forming memories) and disperses throughout the brain.

Alzheimer's is a devastating condition, arguably far worse than Parkinson's, but the good news is that Alzheimer's researchers have recently developed some ingenious tools to help track the disease in the living brain (in fact, more tools than their Parkinson's counterparts). In 2002, for example, the imaging expert Chester Mathis and the psychiatrist William E. Klunk developed a way to detect and measure amyloid-beta deposition in the brains of living Alzheimer's patients. The PET imaging method involves injecting patients intravenously with a radioactively labeled compound that can penetrate the blood-brain barrier and bind selectively to the amyloid-beta protein, wherever it is wreaking damage in the brain. Mathis's new radioactive tracer is called Pittsburgh Compound-B, or PiB (named for the university where Mathis worked). As in other PET imaging techniques, the radiation emitted by the tracer, which preferentially binds to the amyloid-beta deposits, is captured by banks of detectors that surround the patient's head. The data are then converted into a picture that represents the relative amounts of plaque buildup in different parts of the patient's brain. No such imaging technique yet exists that can detect alpha-synuclein aggregates in the brains of people with Parkinson's disease.

PET imaging is an example of a biomarker, a metric that can quantify the progress of a disease in a living patient more sensitively (and, in theory, much earlier) than a clinical test like the UPDRS. Other potential biomarkers are chemical assays that estimate the amyloid content in a patient's cerebrospinal fluid, circulating blood, and urine. So far, there are very few validated biomarkers, but researchers agree that discovering them is a priority because they will help scientists to identify powerful agents that could combat neurodegenerative disease early, much as statin drugs (which lowered the cholesterol biomarker) did for heart disease. As Karl Kieburtz says, "This idea of treating the disease before the illness is something we do in hypertension, something we do with diabetes, something we do in HIV. That is, if your blood pressure is high, we treat you before you have a stroke. If your glucose is high, we treat it before you lose your vision with diabetes." The issue is, can we find such objective, standardized metrics for diseases like Alzheimer's and Parkinson's?

Biomarkers don't ensure success. In the past decade, with the help of the PiB compound and other biomarkers, Alzheimer's researchers have tested numerous drugs designed to target and break down amyloid-beta plaques. While some of these drugs (such as bapineuzumab, solanezumab, Gammagard, and crenezumab) showed promise in open-label studies, they all failed when tested in phase 3 trials in patients with moderate to severe Alzheimer's. The drugs slightly reduced the amyloid-beta burden, but they did little or nothing to arrest the cognitive decline of patients.

One possible reason for this spate of failures, researchers argue, is that the drugs were given too late in the disease; in other words, the patients in the trial might have accumulated too much amyloid beta and developed too much brain damage to respond to these therapies. This argument is analogous to Jeff Kordower's suggestion that gene therapy with growth factors for Parkinson's disease had failed because by four or five years post-diagnosis the damage to the dopaminergic fibers was too extensive.

Alzheimer's researchers, however, have risen to the challenge. They wondered if there might be a way to deliver these plaque-busting drugs much earlier—years before a patient developed any cognitive impairment—and then track any reduction of plaque along with changes in patients' cognition. This meant identifying groups of presymptomatic patients who were virtually certain to develop Alzheimer's disease within five or ten years.

But where do you find an asymptomatic population that is destined or at least highly likely to develop Alzheimer's? Three fascinating early-intervention Alzheimer's studies currently under way claim to have done just that.

The first involves an extended family in South America, and it's a compelling story of scientific detection comparable to the story of the Contursi kindred in Parkinson's disease. About three hundred years ago, a couple from the Basque region of Spain settled in northern Colombia. One of them carried a rare mutation (as we now know, an error located on chromosome 14) for a heritable form of Alzheimer's disease. As this was a dominant mutation, on average, half of this couple's descendants inherited the bad gene, developed signs of mild cognitive impairment in their mid-forties, and progressed to full-blown dementia five or six years later.

By the late twentieth century, the mutation had spread to around five thousand people in a rural area surrounding the city of Medellín. The unique Alzheimer's kindred came to the attention of the University of Antioquia neuroscientist Francisco Lopera. Lopera, who grew up in the town of Yarumal, about a two-hour drive north of Medellín, realized this extended family provided a rare opportunity to test new drugs for Alzheimer's on patients *before* they developed clinical symptoms. Because a simple genetic screening test predicted with 100 percent accuracy which kindred members possessed the bad gene and would contract Alzheimer's disease, these individuals with the "Paisa" mutation (so-called because the people from Antioquia are known as *paisas*) could be given anti-amyloid-beta drugs long before they showed any cognitive impairment. Perhaps then, Lopera and his colleagues reasoned, drugs like Genentech's crenezumab, which had previously failed in phase 3 studies, would prove effective and, like a vaccine, prevent the clinical disease from ever happening.

Lopera's project is part of the Alzheimer's Prevention Initiative headed up by the psychiatrist Eric Reiman at the Banner Alzheimer's Institute in Phoenix, Arizona. In the trial, the Paisa volunteers get a baseline cognitive assessment plus a biomarker evaluation (involving PET imaging, cerebrospinal fluid analysis, and other assays) to measure the distribution of amyloid beta. Then subjects receive either Genentech's drug crenezumab or a placebo. Researchers follow the patients for at least five years. If affected individuals receiving the active drug resist developing mild cognitive impairment, or if that impairment is delayed, then that will be persuasive evidence that the drugs are working—at least in this population. It may serve to validate the biomarkers as well.

A second genetic study—also part of Banner's Alzheimer's Prevention Initiative—involves some thirteen hundred currently healthy individuals aged sixty to seventy-five in Europe and North America. Members of this cohort, however, are at high risk of developing Alzheimer's because they have been identified as carrying two copies of a different gene called apolipoprotein E4. Such people are not guaranteed to develop Alzheimer's like carriers of the Paisa mutation, but their risk is extremely high—about ten times the average probability of getting sporadic Alzheimer's disease. Study subjects will receive either a placebo or one of two experimental drugs developed by the Swiss pharmaceutical company Novartis. Again, if at-risk individuals receiving the active drug resist developing mild cognitive impairment, or if that impairment is delayed, it will be persuasive evidence that the drugs are working in this cohort. Then clinicians would want to know if the same drugs could slow the progress of regular Alzheimer's disease.

The third study that I find intriguing when it comes to its applicability to Parkinson's trials—known as the A4 trial—is supervised by Dr. Reisa Sperling of Harvard Medical School. It involves a cohort of one thousand healthy individuals seventy years of age or older. The individuals in the group exhibit normal cognitive abilities but also have higher than normal levels of amyloid-beta plaques in their brains—as measured by a PET scan. This puts them at higher risk for developing Alzheimer's. The question is, can that risk be reduced or eliminated? Participants will be assigned at random to receive either the experimental amyloid-busting drug or a placebo and will be monitored over the course of three years. As with the other studies, researchers will carry out a full battery of cognitive and biomarker assessments to see if the drugs slow, stop, or even reverse the progression of amyloid plaques. Definitive published results will take time to emerge—at least five years—but there will likely be a steady stream of intermediate discoveries emerging from Alzheimer's researchers over the next few years.

With all of this in mind, what can Parkinson's researchers learn from their colleagues working in the Alzheimer's field? First and foremost, that researchers need smart tools that go beyond clinical tests like the UPDRS. Parkinson's researchers require imaging and other biomarkers for alpha-synuclein that enable scientists to identify people (possibly in the prodromal stage of the condition) and also to monitor the disease as it progresses. Chester Mathis and his colleagues (who pioneered the first imaging test for amyloid beta) are developing a radiotracer that will bind to alpha-synuclein and yield a PET image for Parkinson's. Their effort is part of a consortium set up by the Michael J. Fox Foundation for Parkinson's Research to develop an alpha-synuclein PET tracer as quickly as possible.

The PET tracer is just one element in a bigger Fox Foundation project—the Parkinson's Progression Markers Initiative—designed to discover reliable biomarkers for Parkinson's disease (quite similar to what researchers did for Alzheimer's). The idea is to follow several groups of people forward in time (over several

years), performing clinical examinations, taking biological samples, and doing multiple imaging scans along the way. There's
a cohort of over four hundred recently diagnosed patients not
taking any medication. There's a cohort of around two hundred
without Parkinson's to act as a control group. Then there is a
cohort of around six hundred patients with specific mutations for
Parkinson's disease; these include surviving cases from the famous
Contursi and Iowa kindreds (who have an alpha-synuclein mutation) and also patients with the LRRK2 mutation. And there's
a group of patients without Parkinson's that manifests prodromal
signs like loss of smell and REM sleep behavior disorder. The aim
of the research on all of these groups is to see if any of the markers
(or clusters of markers) reliably track the disease.

Other researchers are looking elsewhere for Parkinson's biomarkers that might allow for early diagnosis. Neuropathologists
had noted finding Lewy bodies and Lewy neurites in diverse locations from the gut to the nose. Inspired by this, Kathleen Shan-
non and her colleagues at Rush University Medical Center accessed
old colon biopsies performed during routine colonoscopy procedures for three individuals who went on to develop Parkinson's
disease. Because these biopsies were done two to five years before
the onset of the Parkinson's motor symptoms, they provided an
opportunity to see if alpha-synuclein was present in the gut before motor symptoms emerged.

It was. The biopsy tissue from the three patients contained lots
of alpha-synuclein, but the biopsy tissue of twenty-three controls,
people who did not develop Parkinson's disease, did not contain
alpha-synuclein. While it's not clear if this provides a viable way
of detecting Parkinson's, it is a provocative idea. Another possible place to look for alpha-synuclein, according to Charles Adler
of the Mayo Clinic in Scottsdale, Arizona, is the submandibular
gland, located just under the jaw. Adler and his colleagues biop-
sied the submandibular glands in twelve people with Parkinson's

for more than five years and found Lewy pathology in nine of them.

The hunt for biomarkers has even attracted the interest of mathematicians. The British mathematician Max Little has developed computer algorithms to analyze human voice recordings to detect irregular patterns in Parkinson's patients. His Parkinson's Voice Initiative (which is supported by the Michael J. Fox Foundation for Parkinson's Research) uses phone call data as potential biomarkers to diagnose and measure the progression of Parkinson's disease.

Patients are hopeful that these multiple efforts inspired in part by researchers in Alzheimer's disease will yield meaningful biomarkers that Parkinson's scientists can use in their research to track the disease in living patients. But none of this will matter unless researchers also come up with potent agents that can significantly reduce the toxic aggregates, especially alpha-synuclein, in the brain. Fortunately, the news on that front is extremely exciting.

15

MEDICINAL GOLD

In 2004, the chemist Chris Dobson speculated that there might be a universal elixir out there that could combat not just alpha-synuclein for Parkinson's but the amyloids caused by many protein-misfolding diseases at once. Remarkably, in that same year an Israeli scientist named Beka Solomon discovered an unlikely candidate for this elixir, a naturally occurring microorganism called a phage.

Professor Solomon of Tel Aviv University made a serendipitous discovery one day when she was testing a new class of agents against Alzheimer's disease. On the face of it, this discovery is potentially far more important than anything I've discussed during the course of this book. If it pans out, it might mark the beginning of the end of Alzheimer's, Parkinson's, and many other neurodegenerative diseases. It's a remarkable story, and the main character isn't Solomon or any other scientist but a humble virus that scientists refer to as M13.*

Among the many varieties of viruses, there is a kind that only infects bacteria. Known as bacteriophages, or just phages, these microbes are ancient (over three billion years old) and ubiquitous:

---

*M13 is tiny, just one micrometer (millionth of a meter) long and six nanometers (billionths of a meter) wide.

they're found everywhere from the ocean floor to human stom-
achs. The phage M13's goal is to infect just one type of bacteria,
Escherichia coli, or *E. coli*, which can be found in copious amounts
in the intestines of mammals. Like other microorganisms, phages
such as M13 have only one purpose: to pass on their genes. In
order to do this, they have developed weapons to enable them to
invade, take over, and even kill their bacterial hosts. Before the
advent of antibiotics, in fact, doctors occasionally used phages to
fight otherwise incurable bacterial infections.*

To understand Solomon's interest in M13 requires a little
background about her research. Solomon is a leading Alzheimer's
researcher, renowned for pioneering so-called immunotherapy
treatments for the disease (this class includes some of the Alzheimer's drugs I mentioned in the previous chapter, such as crenezumab). Immunotherapy employs specially made antibodies,
rather than small molecule drugs, to target the disease's plaques
and tangles. As high school students learn in biology class, antibodies are Y-shaped proteins that are part of the body's natural
defense against infection. These proteins are designed to latch
onto invaders and hold them so that they can be destroyed by the
immune system. But since the 1970s, molecular biologists have
been able to genetically engineer human-made antibodies, fash-
ioned to attack undesirable interlopers like cancer cells. In the
1990s, Solomon set out to prove that such engineered antibodies
could be effective in attacking amyloid-beta plaques in Alzheimer's
as well.

In 2004, she was running an experiment on a group of mice
that had been genetically engineered to develop Alzheimer's disease plaques in their brains. She wanted to see if human-made
antibodies delivered through the animals' nasal passages would
penetrate the blood-brain barrier and dissolve the amyloid-beta

---

*A fictionalized account of a doctor's use of phages to combat an epidemic was depicted in
Arrowsmith, the Pulitzer Prize–winning novel by Sinclair Lewis.

plaques in their brains. Seeking a way to get more antibodies into the brain, she decided to attach them to M13 phages, in the hope that the two acting in concert would better penetrate the blood-brain barrier, dissolve more of the plaques, and improve the symptoms in the mice—as measured by their ability to run mazes and perform similar tasks.

Solomon divided the rodents into three groups. She gave the antibody to one group. The second group got the phage-antibody combination, which she hoped would have an enhanced effect in dissolving the plaques. And as a scientific "control," the third group received the plain phage M13.

Because M13 cannot infect any organism except *E. coli*, she expected that the control group of mice would get absolutely no benefit from the phage. But, surprisingly, the phage *by itself* proved highly effective at dissolving amyloid-beta plaques and in laboratory tests improved the cognition and sense of smell of the mice. She repeated the experiment again and again, and the same thing happened. "The mice showed very nice recovery of their cognitive function," says Solomon. And when Solomon and her team examined the brains of the mice, the plaques had been largely dissolved. She ran the experiment for a year and found that the phage-treated mice had 80 percent fewer plaques than untreated ones. Solomon had no clear idea how a simple phage could dissolve Alzheimer's plaques, but given even a remote chance that she had stumbled across something important, she decided to patent M13's therapeutic properties for the University of Tel Aviv. According to her son Jonathan, she even "joked about launching a new company around the phage called NeuroPhage. But she wasn't really serious about it."

The following year, Jonathan Solomon—who'd just completed more than a decade in Israel's special forces, during which time he got a BS in physics and an MS in electrical engineering—traveled to Boston to enroll at the Harvard Business School. While he studied for his MBA, Jonathan kept thinking about the phage his mother had investigated and its potential to treat terrible diseases like Alzheimer's. At Harvard, he met many brilliant would-be entrepreneurs, including the Swiss-educated Hampus Hillerstrom, who, after studying at the University of St. Gallen near Zurich, had worked for a European biotech venture capital firm called HealthCap.

Following the first year of business school, both students won summer internships: Solomon at the medical device manufacturer Medtronic, and Hillerstrom at the pharmaceutical giant AstraZeneca. But as Hillerstrom recalls, they returned to Harvard wanting more: "We had both spent . . . I would call them 'weird summers' in large companies, and we said to each other, 'Well, we have to do something more dynamic and more interesting.' "

In their second year of the MBA, Solomon and Hillerstrom took a class together in which students were tasked with creating a new company on paper. The class, says Solomon, "was called a field study, and the idea was you explore a technology or a new business idea by yourself while being mentored by a Harvard Business School professor. So, I raised the idea with Hampus of starting a new company around the M13 phage as a class project. At the end of that semester, we developed a mini business plan. And we got on so well that we decided that it was worth a shot to do this for real."

In 2007, with $150,000 in seed money contributed by family members, a new venture, NeuroPhage Pharmaceuticals, was born. After negotiating a license with the University of Tel Aviv to explore M13's therapeutic properties, Solomon and Hillerstrom reached out to investors willing to bet on M13's potential therapeutic powers. By January 2008, they had raised over $7 million and started hiring staff.

Their first employee—NeuroPhage's chief scientific officer— was Richard Fisher, a veteran of five biotech start-ups. Fisher recalls feeling unconvinced when he first heard about the miraculous phage. "But the way it's been in my life is that it's really all about the people, and so first I met Jonathan and Hampus and I really liked them. And I thought that within a year or so we could probably figure out if it was an artifact or whether there was something really to it, but I was extremely skeptical."

Fisher set out to repeat Beka Solomon's mouse experiments and found that with some difficulty he was able to show the M13 phage dissolved amyloid-beta plaques when the phage was delivered through the rodents' nasal passages. Over the next two years, Fisher and his colleagues then discovered something totally unexpected: that the humble M13 virus could also dissolve other amyloid aggregates—the tau tangles found in Alzheimer's and also the amyloid plaques associated with other diseases, including alpha-synuclein (Parkinson's), huntingtin (Huntington's disease), and superoxide dismutase (amyotrophic lateral sclerosis). The phage even worked against the amyloids in prion diseases (a class that includes Creutzfeldt-Jakob disease). Fisher and his colleagues demonstrated this first in test tubes and then in a series of animal experiments. Astonishingly, the simple M13 virus appeared in principle to possess the properties of a "pan therapy," a universal elixir of the kind the chemist Chris Dobson had imagined.

This phage's unique capacity to attack multiple targets attracted new investors in a second round of financing in 2010. Solomon recalls feeling a mix of exuberance and doubt: "We had something interesting that attacks multiple targets, and that was exciting. On the other hand, we had no idea how the phage worked."

That wasn't their only problem. Their therapeutic product, a *live* virus, it turned out, was very difficult to manufacture. It was also not clear how sufficient quantities of viral particles could be delivered to human beings. The methods used in animal experiments—inhaled through the nose or injected directly into the brain—were unacceptable, so the best option available appeared to be a so-called intrathecal injection into the spinal canal. As Hillerstrom says, "It was similar to an epidural; this was the route we had decided to deliver our virus with."

While Solomon and Hillerstrom worried about finding an acceptable route of administration, Fisher spent long hours trying to figure out the phage's underlying mechanism of action. "Why would a phage do this to amyloid fibers? And we really didn't have a very good idea, except that under an electron microscope the phage looked a lot like an amyloid fiber; it had the same dimensions."

Boston is a town with enormous scientific resources. Less than a mile away from NeuroPhage's offices was MIT, a world center of science and technology. In 2010, Fisher recruited Rajaraman Krishnan—an Indian postdoctoral student working in an MIT laboratory devoted to protein misfolding—to investigate the M13 puzzle. Krishnan says he was immediately intrigued. The young scientist set about developing some new biochemical tools to investigate how the virus worked and also devoured the scientific literature about phages. It turned out that scientists knew quite a lot about the lowly M13 phage. Virologists had even created libraries of mutant forms of M13. By running a series of experiments to test which mutants bound to the amyloid and which ones didn't, Krishnan was able to figure out that the phage's special abilities involved a set of proteins displayed on the tip of the virus, called GP3. Says Krishnan, "We tested the different variants for examples of phages with or without tip proteins, and we found that every time we messed around with the tip proteins, it lowered the phage's ability to attach to amyloids."

Virologists, it turned out, had also visualized the phage's structure using X-ray crystallography and nuclear magnetic resonance imaging. Based on this analysis, those microbiologists had predicted that the phage's normal mode of operation in nature was to deploy the tip proteins as molecular keys; the keys in effect enabled the parasite to "unlock" *E. coli* bacteria and inject its DNA. Sometime in 2011, Krishnan became convinced that the phage was doing something similar when it bound to toxic amyloid aggregates. The secret of the phage's extraordinary powers, he surmised, lay entirely in the GP3 protein.

As Fisher notes, this is serendipitous. Just by "sheer luck, M13's keys not only unlock *E. coli*; they also work on clumps of misfolded proteins." The odds of this happening by chance, says Fisher, are very small. As he explains, "Viruses have exquisite specificity in their molecular mechanisms, because they're competing with each other . . . and you need to have everything right, and the two locks need to work exactly the way they are designed. And this one way of getting into bacteria also works for binding to the amyloid plaques that cause many chronic diseases of our day."

Having proved the virus's secret lay in a few proteins at the tip, Fisher, Krishnan, and their colleagues wondered if they could capture the phage's amyloid-busting power in a more patient-friendly medicine that did not have to be delivered by epidural. So over the next two years, NeuroPhage's scientists engineered a new antibody (a so-called fusion protein because it is made up of genetic material from different sources) that displayed the critical GP3 protein on its surface so that, like the phage, it could dissolve amyloid plaques. Fisher hoped this novel manufactured product would stick to toxic aggregates just like the phage. By 2013, NeuroPhage's researchers had tested the new compound, which they called NPT088, in test tubes and in animals, including nonhuman primates. It performed spectacularly, simultaneously targeting multiple misfolded proteins such as amyloid beta, tau, and alpha-synuclein at various stages of amyloid assembly. According to Fisher, NPT088 didn't stick to normally folded individual proteins; it left normal alpha-synuclein alone. It stuck only to misfolded proteins, not just dissolving them directly, but also blocking their prion-like transmission from cell to cell: "It targets small aggregates, those oligomers, which some scientists consider to be toxic. And it targets amyloid fibers that form aggregates. But it doesn't stick to normally folded individual proteins." And as a bonus, it could be delivered by intravenous infusion.

There was a buzz of excitement in the air when I visited Neuro-Phage's offices in Cambridge, Massachusetts, in the summer of 2014. The eighteen staff, including Solomon, Hillerstrom, Fisher, and Krishnan, were hopeful that their new discovery, which they called the general amyloid interaction motif, or GAIM, platform, might change history. A decade after his mother had made her serendipitous discovery, Jonathan Solomon was finalizing a plan to get the product into the clinic. As Solomon says, "We now potentially have a drug that does everything that the phage could do, which can be delivered systemically and is easy to manufacture."

Will it work in humans? While NPT088, being made up of large molecules, is relatively poor at penetrating the blood-brain barrier, the medicine persists in the body for several weeks, and so Fisher estimates that over time enough gets into the brain to effectively take out plaques. The concept is that this antibody could be administered to patients once or twice a month by intravenous infusion for as long as necessary.

NeuroPhage must now navigate the FDA's regulatory system and demonstrate that its product is safe and effective. So far, NPT088 has proved safe in nonhuman primates. But the big test will be the phase 1A trial expected to be under way by early 2016. This first human study proposed is a single-dose trial to look for any adverse effects in healthy volunteers. If all goes well, Neuro-Phage will launch a phase 1B study involving some fifty patients with Alzheimer's to demonstrate proof of the drug's activity. Patients will have their brains imaged at the start to determine the amount of amyloid beta and tau. Then, after taking the drug for six months, they will be reimaged to see if the drug has reduced the aggregates below the baseline.

As Hillerstrom puts it, "If our drug works, we will see it working in this trial. And then we may be able to go straight to phase 2 trials for both Alzheimer's and Parkinson's." There is as yet no imaging test for alpha-synuclein, but because their drug simultaneously lowers amyloid-beta, tau, and alpha-synuclein levels in animals, a successful phase 1B test in Alzheimer's may be acceptable to the FDA. Says Hillerstrom, "In mice, the same drug lowers amyloid beta, tau, and alpha-synuclein; therefore, we can say if we can reduce in humans the tau and amyloid beta, then based on the animal data, we can expect to see a reduction in humans in alpha-synuclein as well."

Along the way, the company will have to prove its GAIM system is superior to the competition. Currently, there are several drug and biotech companies testing products in clinical trials for Alzheimer's disease, against both amyloid beta (Lilly, Pfizer, Novartis, and Genentech) and tau (TauRx), and also corporations with products against alpha-synuclein for Parkinson's disease (AFFiRiS and Prothena/Roche). But Solomon and Hillerstrom think they have two advantages: multi-target flexibility (their product is the only one that can target multiple amyloids at once) and potency (they believe that NPT088 eliminates more toxic aggregates than their competitors' products). Potency is a big issue. PET imaging has shown that existing Alzheimer's drugs like crenezumab reduce amyloid loads only modestly, by around 10 percent. "One weakness of existing products," says Solomon, "is that they tend to only prevent new aggregates. You need a product potent enough to dissolve existing aggregates as well. You need a potent product because there's a lot of pathology in the brain and a relatively short space of time in which to treat it."

.

NeuroPhage's rise is an extraordinary example of scientific en-
trepreneurship. While I am rooting for Solomon, Hillerstrom, and
their colleagues, and would be happy to volunteer for one of
their trials, there are still many reasons why NeuroPhage has a
challenging road ahead. Biotech is a brutally risky business. At
the end of the day, NPT088 may prove unsafe. And it may still not
be potent enough. Even if NPT088 significantly reduces amyloid
beta, tau, and alpha-synuclein, it's possible that this may not lead
to measurable clinical benefits in human patients, as it has done in
animal models. But if it works, then, according to Solomon, this
medicine will indeed change the world: "A single compound that
effectively treats Alzheimer's and Parkinson's could be a twenty-
billion-dollar-a-year blockbuster drug." And in the future, a modified version might also work for Huntington's, ALS, prion diseases
like Creutzfeldt-Jakob disease, and more.

I asked Jonathan about his mother, who launched this remarkable story in 2004. According to him, she has gone on to other
things. "My mother, Beka Solomon, remains a true scientist. Having made the exciting scientific discovery, she was happy to leave
the less interesting stuff—the engineering and marketing things
for bringing it to the clinic—to us. She is off looking for the next
big discovery."

16

BRAIN STORMS

The war against Parkinson's is a winnable war, and I have
resolved to play a role in that victory.　　—Michael J. Fox

If we could just convince scientists to talk about all the
good things rather than to start shouting each other down
that would be great . . . We need that "can-do" attitude.
　　　—Tom Isaacs, cofounder of Cure Parkinson's Trust

In October 2014, the World Parkinson Congress organized a
series of Webcasts in which leading neuroscientists updated
the Parkinson's community on the latest research developments since the Montreal meeting twelve months before. It had
been a busy year for me as well. With neuroscientists and other
people with Parkinson's as my guides, I had been on a fascinating
journey delving into the past, present, and possible future of Parkinson's research. The experience has left me with a deep respect
for the tenacity of researchers and clinicians and an admiration for
the courage and ingenuity of patients.

On that October day, I realized that the story of Parkinson's
is a case study in the extraordinary complexity and slow pace of

biomedical science. After all, it's taken more than two centuries for a series of brilliant scientists to painstakingly construct a (still incomplete) picture of this neurodegenerative disease. There's James Parkinson, whom we can thank for his astute observations of patients in his London neighborhood and the resulting *Essay on the Shaking Palsy*. There's Jean-Martin Charcot, the perceptive French clinician who nailed the clinical signs and symptoms of what appeared to be a movement disorder. There's Constantin Tretiakoff, the Russian student who demonstrated that the disease was caused by damage to the substantia nigra region of the brain, and there's Frederick Lewy, after whom Lewy bodies, the pathological hallmarks of Parkinson's, are named.

In this narrative, we've also met remarkable scientists such as Walther Birkmayer, Oleh Hornykiewicz, George Cotzias, and Roger Duvoisin who, among others, showed that levodopa could temporarily reverse parkinsonian symptoms in humans. We've encountered Bill Langston, who discovered a neurotoxin, MPTP, in a street drug, which led to a new animal model of the disease; and Mahlon DeLong and Alim-Louis Benabid, who paved the way for deep brain stimulation. Our story includes pioneering researchers like Patrik Brundin and Jeff Kordower, who attempted to protect, revive, or replace damaged dopamine neurons, and Heiko Braak and George Webster Ross, who demonstrated that (as many neurologists and neuropathologists already suspected) Parkinson's was much more than a motor disorder focused on the substantia nigra. We can thank Larry Golbe, Bob Nussbaum, Maria Spillantini, and many others for demonstrating the potentially critical role of alpha-synuclein in Parkinson's pathology. And the field is now poised to test a series of exciting agents designed to stop the spread of this rogue protein in our bodies and brains.

And as the Webcast sessions showed, research continues on many fronts. The cyber presentations covered a wide range of scientific research and clinical topics, addressing both long-term research goals and current clinical care. But four topic areas in particular got my attention as being especially relevant to patients like me—improved L-dopa delivery, the placebo effect, the importance of non-motor symptoms, and the drive to develop personalized medicine. Each one of them raises important outstanding issues for the Parkinson's community.

For nearly fifty years, researchers have sought better ways to deliver levodopa in a manner that would minimize the troublesome motor complications that virtually every person with Parkinson's eventually encounters. But so far, they haven't been able to pull it off. Why have scientists failed to solve this puzzle? As Peter LeWitt, professor of neurology at Wayne State University School of Medicine, explained to me, getting L-dopa into the brain is a complex problem in medicinal chemistry that we're just beginning to figure out. According to LeWitt, here's what happens when a patient swallows a standard 25/100 carbidopa-levodopa tablet: The pill passes down the esophagus and enters the stomach. Within about fifteen to thirty minutes, it reaches the upper gastrointestinal tract—a small eighteen-inch section of intestine that includes the duodenum and the jejunum. This relatively short section is a key departure terminal for molecules destined for the brain. Molecules like levodopa—which are chemically identical to naturally occurring substances in our bodies such as amino acids—are readily absorbed along this short stretch of gut. They are taken up by the large neutral amino acid transporter, a kind of conveyor belt that carries a vast cargo of chemicals exiting the stomach en route to the body's tissues. The 100 milligrams of levodopa in the pill vie for space on this conveyor belt with whatever else is exiting the stomach—such as the amino acids digested from proteins that were part of your breakfast.

As the conveyor belt passes through the body—visiting skeletal muscle, liver, kidneys, and more—enzymes attack its molecular cargo.* These enzymes prematurely convert most of the levodopa into dopamine, which can't cross the blood-brain barrier. So, in an attempt to limit the loss, manufacturers wrap the 100 milligrams of levodopa with 25 milligrams of powerful decarboxylase enzyme blockers (carbidopa or benserazide).

Even with the protection of the decarboxylase inhibitor, says LeWitt, only about 5 to 10 percent of the levodopa in the pill actually makes it to the brain, crosses the blood-brain barrier, and is converted into dopamine. And that dopamine is delivered not steadily to the neuronal receptors but unevenly, starting with small amounts, building to a maximum, and falling to almost nothing over the course of about three hours. As we've seen, people taking the medicine report experiencing this rise and fall, detecting when their medicine kicks in, when it peaks, and when it wears off. As the disease progresses, the patient might have an increasingly difficult time coping with this uneven delivery. And just as we saw with cases like Nancy Egan's, people with advanced Parkinson's end up having to take six or seven pills a day just to keep moving.

Over the years, neuroscientists and pharmaceutical companies have explored various strategies to deliver L-dopa more evenly and continuously to the brain. They made a version of carbidopa-levodopa that is released more slowly in the gastrointestinal tract (Sinemet CR). They tried combining carbidopa-levodopa with another enzyme inhibitor called entacapone (Stalevo). They tried blocking the breakdown of dopamine in the brain (using selegiline and rasagiline), and they even attempted to replace L-dopa with drugs called dopaminergic agonists (pramipexole, ropinirole, and rotigotine), molecules designed to mimic dopamine. None of these significantly improved the patient experience.

LeWitt believes that the efforts failed because the products

---

*The two most important enzymes are aromatic L-amino acid decarboxylase (AADC) and catechol-O-methyltransferase (COMT). Something like 70 percent of the levodopa will be broken down by AADC and another 10 percent by COMT.

were based on "a lot of halfhearted pharmaceutical development, with poor design, and not enough clinical research." Take
Sinemet CR, for example. Once it passed beyond the critical eighteen inches of upper gastrointestinal tract where the absorption
took place, its slowly releasing drug load was simply wasted. Stalevo might have been effective if the entacapone enzyme blocker
had been delivered to the gastrointestinal tract before levodopa.
It wasn't. In studies, Stalevo reduced "off" time but increased dyskinesias. Dopaminergic agonists, for most patients, turned out to
be pale imitations of levodopa and had their own unpleasant side
effects. Monoamine oxidase inhibitors like selegiline and rasagiline had a mild therapeutic effect but were hardly game chang-
ers. However, as LeWitt told me, "When they discovered such
problems, the drug companies didn't drop the project and then
go on to something better; they finished up marketing the product regardless of what the data showed."

Fortunately, not everyone has abandoned the quest for continuous dopamine delivery. Indeed, the Michael J. Fox Foundation for Parkinson's Research has spent millions supporting
several innovative symptom-modifying ideas.* The California-
based company Depomed, for example, embeds the carbidopa-
levodopa in a special polymer designed to swell up on contact with
the stomach's gastric juice. Rather than passing into the ileum and
large intestine, the swollen polymer lodges in the stomach, strate-
gically positioned just above the critical absorbing region of the
upper gastrointestinal tract. Over eight to ten hours the drug-
polymer hybrid erodes, delivering a steady dose of levodopa.
This ingenious idea (a "gastric retention" platform called Acu-
form) has already been approved by the FDA to deliver antibiotics,

---

*In addition to Depomed's and NeuroDerm's technologies (discussed here), the Fox Foun-
dation has supported Civitas, which produces an inhalable levodopa fast-acting "rescue"
therapy, Intec Pharma's "accordion pill" (another gastric-retention strategy for more
continuous dopaminergic effect), and Cynapsus, which makes an oral apomorphine rescue
technology.

diabetes drugs, and pain medications, and the evidence is that it can do the same for L-dopa. In phase 1 and phase 2 trials, patients receiving two daily doses of Depomed's product displayed much more constant levodopa concentrations—avoiding the peaks and troughs—and experienced fewer motor complications than patients getting five doses of immediate-release carbidopa-levodopa.

An Israeli company, NeuroDerm, has developed an even more revolutionary approach—to bypass the gastrointestinal tract altogether and deliver a liquid formulation of carbidopa-levodopa subcutaneously. In a brilliant piece of chemistry, Neuro-Derm scientists figured out how to produce a stable, concentrated *liquid* form of carbidopa-levodopa that can be practically delivered into the bloodstream—a challenge that had defeated chemists for half a century. Using a "pump-patch," or belt pump, similar to those used by type 1 diabetes patients, NeuroDerm's platform delivers carbidopa-levodopa continuously over twenty-four hours.

While such technologies are works in progress,* it's all very exciting. Peter LeWitt says, "The near-constant levels produced by these products are remarkable . . . and comparable with expensive, invasive technologies like Duodopa." (Duodopa, a technology that continuously infuses carbidopa-levodopa gel through a surgically fitted tube in the jejunum, is approved in Europe and Canada but not yet in the United States.)

If such products offer simple, noninvasive ways of achieving the long-sought-after goal of steady L-dopa delivery, why haven't we heard more about them? One disturbing possibility appears to be that despite their potential importance to patients, big pharma finds them commercially unattractive. To get these products to market involves not only phase 1 and 2 studies (which the compa-

---

*Other companies developing dopamine-delivery products include Intec Pharma and XenoPort.

nies have undertaken) but also two expensive phase 3 clinical trials and a one-year-long FDA New Drug Application review. So, typically, small companies like Depomed and NeuroDerm need financial investment, often in the form of a pharmaceutical industry partner, to afford this final stage.

I asked Jerry Callahan, vice president for business development at Van Andel Institute and the founder of two venture capital firms specializing in the life sciences, why giants like Pfizer, GlaxoSmithKline, and Roche aren't lining up to partner with Depomed and NeuroDerm. Because levodopa is an established drug taken by millions of people, the risks of adverse effects are likely vanishingly small. But what about commercial risks? What he told me was sobering. When deciding on investments in the biotech industry, pharmaceutical companies typically ask several questions: Is there an unmet need? Does the new product meet the unmet need? How big is the potential market? Will Medicare, Medicaid, and private insurance companies agree to premium prices?

A life-threatening condition, like an incurable form of cancer or heart disease, is an unmet need for vast numbers of people. A new cancer drug that extends life can probably, therefore, command a premium price from payers. Novartis, for example, charges $76,000 a year for its drug Gleevec. But, says Callahan, "Parkinson's is not like cancer; it's not like heart disease. Patients don't typically die from Parkinson's disease. A person with Parkinson's disease is usually in their sixties, has lived a long life, and many are already retired. It's bad, but it's not life threatening."

The commercial risk analysis goes like this. Because the proposed new products won't extend lives but will rather simply improve the lives of people who already have access to L-dopa, payers like Medicare might not approve premium pricing (say, $10,000 a year compared with the current $1,000 per year spent on Sinemet) for this "convenience." "Now, people who know Parkinson's disease," says the Michael J. Fox Foundation CEO,

Todd Sherer, "know that it's hardly a matter of convenience. Advanced patients need improved delivery because they keep going on and off, and they keep freezing up, and they're dyskinetic . . . A steady, continuous dose of L-dopa would allow people to be more productive, to keep their jobs, creating real-life impacts." Over and above the help it would provide to advanced patients with few options, some neuroscientists believe that in theory treating early-stage cases from the outset with continuous dopamine delivery might avoid motor complications altogether. Such a prospect, if it turns out to be effective, could radically improve a newly diagnosed patient's trajectory.

Researchers, clinicians, and people with Parkinson's need to keep up the pressure on the pharmaceutical industry so that good ideas, like those related to levodopa delivery, make it to the clinic. In the future, something as apparently "low tech" as a better delivery system for an old drug might greatly improve the lives of people with Parkinson's. And after fifty years, it's about time.

On a different note, the Rush University neuroscientist Christopher Goetz mentioned in an update to the Parkinson's community the intriguing and somewhat controversial topic of the placebo effect, which I talked about in chapter 8. Goetz the clinical neurologist believes it is an effect worth keeping. As he puts it, "I use the placebo effect when I greet my patients, when I encourage them, when I tell them we're a partnership . . . [I] would never want to eliminate it in the clinic." But Goetz the scientist sees the placebo effect as a liability. "In a trial, if the patient gets just as good effect with sham surgery as having some kind of foreign cell implanted, then we have a problem." That's the conundrum in a nutshell.

In 2008, Goetz carried out an ingenious study to unpack the true dimensions of the placebo effect as it relates to Parkinson's. He contacted the investigators of eleven major medical and surgical Parkinson's trials and procured information about the patients in the placebo arms (858 subjects in total), patients who thought they had received the test drug or surgery but who had in fact received a placebo.

In every one of these studies, neurologists had evaluated patients' motor symptoms at the start of the studies (the "baseline") and then at regular intervals throughout the follow-up period of around six months. The clinicians used the ubiquitous Unified Parkinson's Disease Rating Scale motor subscale (UPDRSm); as we've mentioned, this is a clinical battery of tests, such as hand flips and finger taps, that are scored on a four-point scale (0, normal; 1, slight; 2, mild; 3, moderate; or 4, severe). On this scale, a higher aggregate score means more severe parkinsonism.

When taking a proven therapy like L-dopa, patients perform the tests better, and the aggregate UPDRS total falls (in other words, it improves) on average by about one-third. Goetz decided he would apply an even higher bar for the placebo effect. Only if a patient in the placebo group improved his score by *half* would Goetz count him as a placebo responder. As he told me, "I wanted something that was incontestable, something that if you saw that kind of change, you would say 'wow.'"

His analysis found there was indeed a subset of patients whose UPDRS dropped by half when they took the placebo therapy. Across all eleven studies, an average of 16 percent were such "placebo responders," but in two of the surgical studies the proportion was much higher—over half of patients getting sham surgery experienced a strong placebo response. Remarkably, this robust placebo effect persisted throughout the six-month follow-up period. In one study transplanting fetal material dissected from the brains of pig fetuses, the placebo group was still showing markedly lower UPDRS numbers more than eighteen months after surgery.

Goetz had found evidence of a strong placebo effect (equal or greater to the therapeutic power of L-dopa) that in some patients persisted for many months. This has profound implications. After all, Parkinson's disease is a progressive condition that inexorably worsens over time. Says Goetz, this "continuing deterioration should mean the scores get worse, but in fact the scores get better. That's what's so remarkable. [The placebo effect] is overwhelming the disease progression."

What are we to make of such a finding? Goetz's gut feeling is that the placebo's got a lot to do with doctors and clinical settings. "It's very typical that when a patient comes into the office for an appointment, they look—according to the spouse—30 to 40 percent better than usual . . . just through coming to the doctor!" The University of Tübingen's Walter Maetzler agrees, saying that "heavy freezers"—patients with pervasive freezing of gait—sometimes "do not develop freezing when they go into the doctor's office. And the wife is complaining, 'I don't understand what is going on; at home, he isn't able to perform anything.' And she turns to the husband and asks, 'Are you faking it at home?'" There are various theories to explain this paradoxical movement. Maetzler speculates that when freezers are a little bit nervous, they increase their level of attention, and that in turn might use different brain circuits that somehow allow them to bypass the disability.

Just why do clinical settings seem to matter so much? It could have something to do with what placebo researchers call the "ritual of the therapeutic act." Even without a pharmacologically active medicine, a person in a clinic or an operating room, they argue, experiences a complex psychosocial process. A patient arrives at the medical center with expectations based on her knowledge, beliefs, and memories of past medical encounters. All of her senses are involved in the subsequent interaction. She sees doctors and nurses and observes, smells, and tastes medicines of different forms, colors, and shapes. She feels the touch of instruments like syringes. She hears the beeps and gurgles of medical machinery and listens to the words of those present. This elaborate therapeutic ritual, they argue, is medically potent even if only fake medicines are being delivered.

This leads to a striking conclusion. While the medicines may be fake, the placebo responses going on in our brains must be real, mediated by actual neurotransmitters. As Goetz says, "It's not mind over matter . . . it's coming from dopamine."

Dopamine is known to regulate reward-seeking and novelty-seeking behaviors. So it's not hard to imagine how a dopamine-mediated placebo effect might trigger relief from pain in a healthy patient. But in Parkinson's, it can't be that simple. Parkinson's is a disease defined by dopamine *depletion*. When symptoms first appear, perhaps 70 percent of the dopamine neurons in the substantia nigra (and an even higher proportion of the connections to the striatum) are already gone. Given this major loss, where does the brain find enough dopamine to modulate a placebo effect strong enough to cause major symptomatic improvement that can persist for years?

In 2001, the University of British Columbia researchers Raúl de la Fuente-Fernández and Jon Stoessl conducted the first brain imaging study of the placebo effect using positron-emission tomography. In an ingenious design, patients in the test group were variously injected with apomorphine, a powerful fast-acting drug with L-dopa-type effects, and saline solution (a placebo), but they were not told which was which. Patients in a second "open" comparison group got the same injections, but the researchers told them what they were receiving. Placebo theory suggests that some of those patients expecting apomorphine (but receiving saline) would generate dopamine, whereas those who knew they were getting saline would not.

Testing the theory means looking into the brain, something that a PET scanner can do with the help of radioactivity. Stoessl and Fuente-Fernández intravenously delivered raclopride (a drug that attaches to dopamine receptors in the striatum and, therefore, competes with any dopamine generated inside the patient's own body). Because the raclopride is labeled with a radioactive form of carbon (C11), which emits positrons, researchers can track its passage through the brain. Over the course of an hour, detectors assemble an image of raclopride accumulation in the subject's striatum.

If the patient expecting apomorphine but getting placebo saline actually produces dopamine, then that dopamine will fight with the raclopride for space in the striatum. And the resulting PET image should be smaller. That's exactly what happened. The placebo group all registered a large burst of dopamine in the striatum—comparable, Stoessl says, to "therapeutic doses of levodopa." The members of the open control group (who knew they were getting saline) produced no such dopamine burst.

Other studies have replicated this finding, and the Italian investigator Fabrizio Benedetti has even shown that neurons change their firing rate when patients expect they are getting a hit of dopamine. It seems to be true.

Yet where is the dopamine coming from? This is one of the remaining mysteries of Parkinson's that cries out for an answer. Stoessl, a jovial, avuncular figure, suggests some possible answers. "There could be some surviving dopamine neurons in the nigrostriatal pathway . . . Another possibility is that dopamine is released from other fibers—it's known that serotonergic neurons can release dopamine." He mentions a third possibility as well—that a dopamine-rich region of the midbrain that's nearby (called the ventral tegmental area) might sprout terminals that then release dopamine into the striatum. But he admits it's all speculation at this point.

Whereas researchers are wary of the placebo effect, I view it as a positive and encouraging phenomenon. The brain, it seems, can work like a compounding pharmacy and synthesize chemicals like opiates, dopamine, and who knows what else. Finding ways to harness this ability (as happens intermittently when a patient gives an improved performance in the clinic) might offer exciting new strategies for patients to optimize their condition.

.

While dopamine—sourced from L-dopa therapy or placebo expectations—remains critically important for Parkinson's therapy, the scientific consensus now holds, as I've discovered during the course of my investigations and research on my own diagnosis, that dopamine is only part of the story. It's now accepted that the underlying pathology extends way beyond the substantia nigra, affecting non-dopaminergic networks in the brain as well, like the serotonin, epinephrine, and acetylcholine systems. And over time, this neuronal damage leads people with Parkinson's to develop a long list of non-motor symptoms (from insomnia to fatigue, from excessive sweating to sudden drops in blood pressure, from loss of taste to double vision) in addition to the motor problems they are already grappling with. The British neuroscientist Ray Chaudhuri was among those whose Webcast reported on the challenges of treating Parkinson's patients in an increasingly non-dopaminergic world. Researchers, clinicians, and patients are still digesting this profound rebranding of Parkinson's. Most patients find this new picture of Parkinson's disease somewhat discomforting for one main reason: many of the most debilitating non-motor symptoms concern the loss of mental abilities in one form or another. In fact, because Parkinson's patients suffer from so many psychiatric symptoms, the psychiatrist Daniel Weintraub has even called Parkinson's "the quintessential neuropsychiatric illness."

As Ray Chaudhuri has argued, people with Parkinson's can expect to confront an alarming set of neuropsychiatric conditions. There's depression and anxiety, which affect about half of all Parkinson's sufferers. Depression and anxiety can even present in the prodromal phase—in other words, before diagnosis. My family and friends have told me that on occasion, I seemed a little uncharacteristically depressed in the years before my diagnosis.

There's also a high prevalence of sleep disorders among people with Parkinson's—including REM sleep behavior disorder, excessive daytime sleepiness (which Chaudhuri likens to narcolepsy), restless legs syndrome, and insomnia. Chaudhuri also spoke about the neglected area of Parkinson's-related pain, which he argued included motor pain, central pain, and unexplained lower-leg pain.

Other common symptoms include apathy (defined as lack of motivation, with a decrease in goal-directed behavior and flattened affect) and hallucinations. But perhaps worst of all, especially for people who have made their living using their minds, is the prospect of impaired cognition and dementia.

Whereas Alzheimer's disease affects memory, in Parkinson's cognitive deficits tend to affect a person's "executive function." This is a critical system designed to manage other cognitive processes, a crucial part of our mental vitality that includes the regulation of working memory, problem solving, the capacity to plan goals and execute actions, and more. In a riveting *New Yorker* article, the journalist Michael Kinsley, who has had Parkinson's for twenty years, courageously puts his brain to the test to see what he's lost. The resulting report on Kinsley's mental function, written by the psychologist Mark Mapstone, starts out quite well, containing such comments as "Mr. Kinsley is a highly intelligent, friendly, and engaging 62-year-old man . . . IQ in the Very Superior Range . . . excellent cognitive reserve . . . exceptionally strong vocabulary." But then, says Kinsley, "when they get to 'executive functioning,' the whole thing turns south. It begins: 'Mr. Kinsley's performance varied significantly across measures of executive functioning. Executive dysfunction . . . included poor organizational skills, weak verbal fluency, inefficient problem solving, a tendency to break task rules, and weak working memory.'"

For people like Kinsley and myself, who have always relied on our intellect rather than our physical strength and dexterity, the idea that Parkinson's will rob us of mental acuity along with motor function is very disturbing. Even more threatening is the notion that cognitive impairment can progress into full-blown dementia. Some longitudinal studies indicate that up to 80 percent of Parkinson's patients will eventually lose the ability to think and reason. The main risk factor for dementia seems to be a patient's current age, not the duration of his disease. So it seems the Parkinson's patient is in a trap. The longer you live, the greater your chance of losing your mind.

Kinsley argues not only that this reality is unpleasant for the individual suffering from Parkinson's but also that there are broader societal issues at stake. People diagnosed with Parkinson's, he warns, might face discrimination of many kinds, especially in the workplace. As he puts it, "As the word gets out that Parkinson's disease is not just a movement disorder, there will be people whose careers will be destroyed because, on a particular day at a particular time, they can't [do a test asking them to] recite a seven-digit telephone number backward."

Kinsley raises some important issues. Unfortunately, I have found few people with Parkinson's who are very keen to talk about the possibilities of cognitive impairment and dementia. For many people, it may be easier not to address them. One way to approach the topic, however, may come from another big theme in Parkinson's research: the idea of personalized care.

As we know, Parkinson's is a highly variable disease. It presents at different ages, with different clinical subtypes that progress at different rates, in people facing widely different socioeconomic circumstances. And because everybody's Parkinson's disease is different, clinicians are increasingly arguing that the care of people with Parkinson's needs to be "personalized" as much as possible.

The neuroscientists Jon Stoessl, Ronald Pfeiffer, and Roger Barker argued in their Webcast that the age of disease onset is a huge factor in determining the best treatment. As we age, our brains and bodies undergo profound changes. Our brains lose their plasticity, and some of our vital metabolic processes slow down in the liver, kidneys, and other major organs. This means that some drugs that might work in younger patients may cause side effects in older patients. For example, anticholinergic drugs like Artane and Cogane, which block the neurotransmitter acetylcholine, thereby relieving a range of symptoms (muscle stiffness, sweating, excess production of saliva, and more), present risks of side effects such as confusion and delirium primarily in people over the age of sixty. And dopamine agonists like Mirapex may work well in younger patients but cause psychiatric problems and hallucinations in people over seventy.

Life circumstances may also affect a clinician and patient's choice of drugs. A person with Parkinson's working in a job where she has to perform before an audience, say, as a TV anchor or teacher, may want to more aggressively treat her tremor than someone working in a less public position. Likewise, people with Parkinson's who have jobs that require them to carry out fine-motor tasks with dexterity may seek different medication regimens than those who are retired.

And as Stoessl emphasized, it's not just a question of drugs. How a person with Parkinson's chooses to live may be as crucial to his well-being as which medicines he takes. Research supports the idea that patients who exercise regularly (as I explored in chapter 7) and who keep a positive attitude and remain socially and mentally engaged do much better than those who withdraw from the world. Whether this is because of the neuroprotective effects of exercise and engagement or a robust placebo effect is still to be determined.

We are also seeing evidence that it may be desirable to personalize clinical trials as well as clinical treatments. Charcot had found that one in five Parkinson's patients do not present with a tremor. There is some evidence that patients with Parkinson's who do—the so-called tremor-dominant cases—progress more slowly than patients whose disease is dominated by axial symptoms, such as postural instability and gait. Detecting such clinical subtypes in advance offers neuroscientists valuable information about what treatments are most likely to work on particular patients.

The Cambridge neuroscientist Roger Barker, who has tracked a Parkinson's cohort in England for fifteen years, has reported finding other profound differences in clinical progression between patients. According to Barker, the research reveals two critical subtypes of Parkinson's disease—an older group that tends to develop dementia within a decade and a younger group whose Parkinson's disease has a more benign course. It turns out that the more rapidly progressing group can be predicted ahead of time genetically and clinically.

Genetically, some people in this group inherit a mutation associated with a rare condition called Gaucher's disease. The patient who inherits a Gaucher's mutation doesn't look any different from a regular Parkinson's patient at presentation, but every one of them is demented within eight years in Barker's study.

A remarkably straightforward clinical test is also predictive. According to Barker, "If you can't pass some simple cognitive tests—to produce a list of twenty animals in under ninety seconds and draw two interlocking pentagons—the facts are pretty clear: your chance of dementia after five years is much higher than younger people who can do these tasks." Says Barker, such advanced cases are not suitable for treatments based around new dopamine therapies like neural grafting and neurotrophic factors. But if such cases can be detected early enough, they may be ideal for future disease-modifying therapy trials, studies designed to test agents that might prevent them progressing into dementia.

A disease-modifying therapy is the holy grail of Parkinson's research for one important reason: such a therapy can in principle benefit just about everyone.

Consider a new therapy targeted at toxic forms of alpha-synuclein, for example. For the patient with advanced motor impairment with intact cognition, such a therapy could in principle block the spread of rogue alpha-synuclein to the cortex, preventing dementia. Most people with Parkinson's would gladly live with a disabled body if they could hold on to their minds.

For patients like me with mild motor issues, such a therapy could (by stopping further spread of misfolded alpha-synuclein) give my brain time to adapt so as to more optimally use my remaining set of dopamine, serotonin, epinephrine, acetylcholine, and other neurons. I would happily accept the impairment I have now if I believed it would not get any worse.

For the patient with early prodromal indicators of Parkinson's like constipation and REM sleep behavior disorder, there is also hope. I can imagine such an individual one of these days being treated with an effective immunotherapy and never progressing to a diagnosis of Parkinson's. And ultimately, of course, neuroscientists may one day develop a medicine that prevents the spread of rogue alpha-synuclein at the outset, eliminating even the prodromal symptoms.

It was with such positive thoughts in mind that I set out in late September 2014 for Grand Rapids, Michigan, to attend a conference called "Rallying to the Challenge," a meeting that brought together neuroscientists and people with Parkinson's. The gathering was the brainchild of two remarkable individuals: the Swedish neuroscientist Patrik Brundin and the patient Tom Isaacs, the cofounder of the Cure Parkinson's Trust—the man who had walked around the coastline of Great Britain. Brundin had dedicated his life to finding a cure for the disease that claimed his father's life. Isaacs had devoted all his energies to supporting that quest through disease advocacy. Now twenty years into his

Parkinson's, Isaacs had become convinced that patients needed to play a more integral role in the search for a cure. And Brundin agreed. One area in particular needed to be addressed—persuading enough patients to volunteer for clinical trials designed to find new disease- and symptom-modifying therapies. This turns out to be a major problem facing medicine in general and not just Parkinson's disease. Around 80 percent of all clinical trials get delayed because of difficulties recruiting suitable subjects. Isaacs and Brundin had decided to tackle this problem head-on.

Tom Isaacs presented a bizarre sight as he stumbled his way onto the Van Andel Institute stage to address an audience of 250 scientists and 100 patients. Despite valiant attempts to control his marked dyskinesia, Tom walked, as he would be the first to admit, in a Monty Python "Ministry of Silly Walks" fashion. And when he arrived at the podium and began to talk, his upper body gyrated back and forth so energetically that I feared he would fall over. But Tom stayed on his feet and delivered a truly moving presentation, revealing himself as a great leader. His message was upbeat. "The world of Parkinson's," he said, "is an exciting place to be at the moment, even if you've had it for twenty years like me."

His talk, directed at both patients and researchers, concerned the importance of maintaining hope and urgency. As he put it, at the moment a person is diagnosed with Parkinson's disease, "the sand in the hourglass starts to run out." The challenge as time passes, he argued, is to maintain a positive attitude. "So much of this illness is about mind-set. People with Parkinson's disease here at this meeting don't just live with this condition; they positively defy it." But, he continued, "if we want to lead better lives, there needs to be rapid progress, bringing new treatments to market. The only way to do this is through clinical trials." Then, addressing the scientists in the audience, he announced,

"That's why we're here. We want to help you so that you can help us."

Over the next thirty-six hours, scientists and patients began drawing up a new protocol that would involve people with Parkinson's in the design, recruitment, and execution of future clinical trials. A whole range of issues were discussed—from the need for improved communication to the necessity of reimbursing patients for travel costs, from the insistence that patients assigned to the placebo arm of a trial receive the active treatment at the end of a successful clinical trial to the suggestion that some patients would be willing to take more risk (due to the severity of their condition) than others. While these conversations were just the first steps in a long process, I detected a feeling of great excitement among the delegates.

For me, it was inspiring to see so many brave and engaged people with Parkinson's come together with researchers who are devoting their professional lives to conquering our disease. Among the patients, there was a six-foot-nine-inch former NBA player named Brian Grant. Grant, whose career involved playing for the Sacramento Kings, the Portland Trail Blazers, the Miami Heat, the Los Angeles Lakers, and the Phoenix Suns, was diagnosed with Parkinson's at the age of thirty-six. At first, he says, he was embarrassed by his tremor, but he soon learned to "lose the vanity, because when you have this disease, you better learn how to laugh at it, because otherwise you're going to be sitting around pissed off with it all the time." Grant, like so many other people with Parkinson's, drew on his inner resources to cope with his new life. As he told us, "As human beings, we each have our stuff that we have to deal with . . . The athlete in me serves me well in this challenge, because there are days when I feel depressed, there are days when I just don't feel like doing it. It's kind of like practicing for Coach Pat Riley: you've just got to do it. So with Parkinson's, I got to get my butt out; I got to get to the gym." Good advice indeed.

·

At the end of the first day of the meeting, we assembled in downtown Grand Rapids for a rather cool celebration of the Parkinson's community. Instead of staging a flash mob, we decided to perform a flash freeze. At precisely 6:00 p.m., a group of some seventy people with Parkinson's (plus a few neuroscientists), dressed in white T-shirts with the words "Still Life" written in red, performed a synchronized two-minute freeze. The words "Still Life" were chosen for their double meaning—indicating both the lack of movement afflicting people with Parkinson's and the idea that people with Parkinson's still have a lot of life left in them. We all froze like statues—some of us mid-stride as if running or walking, others holding an exaggerated hand gesture, still others looking at another person as if frozen in a conversation. Passersby didn't know quite what to make of us. During those two minutes of shared frozen silence, I thought about the journey I was on and took comfort in the fact that I wasn't making it alone.

Is it fair to say that we are on the verge of winning the war on Parkinson's disease? The patient in me would like to think so. But on several occasions in the past, wishful predictions that a cure was imminent have turned out to be misplaced. Recall that in 1999 the then NINDS director, Gerald Fischbach, stated, "We are close to solving—and I mean the word 'solving'—Parkinson's disease." And remember how Michael J. Fox told Barbara Walters, "There are so many things on the horizon. So many medications, and surgical procedures . . . I really feel that within the next years they're going to find a way to flick a switch, and this is gone."

Today, those statements seem courageous but naive. On the other hand, science has moved on. And now so much promising research is going on around the world that it's reasonable to believe that sooner or later victory will be ours. Following in the footsteps of the Human Genome Project, biomedical scientists have recently launched a series of epic brain-mapping quests. On April 2, 2013, the Obama administration announced the BRAIN Initiative; BRAIN stands for Brain Research Through Advancing Innovative Neurotechnologies. The $300 million, ten-year project seeks to map the activity of every neuron in the human brain. A related enterprise, the Human Connectome Project, plans to build a "network map" that will compare how neurons are anatomically and functionally connected in healthy and diseased brains. And in 2013, European scientists embarked on the ten-year Human Brain Project, designed to simulate the complete human brain on powerful supercomputers—eventually fashioning a full computer model of a functioning brain that might be used to test potential drug therapies. Such grand projects can be expected to shake up neuroscience every bit as much as the Human Genome Project transformed human genetics.

In my case, I will be watching closely as NeuroPhage and other companies test their amyloid-busting products. If the safety data look good, I will volunteer for a trial. A disease-modifying therapy that allowed me to hold on to what I have would obviously be much better than a continued decline. And even if those disease-modifying therapies fail to pan out, I will be tracking the companies like Depomed and NeuroDerm that are trying to develop products that deliver L-dopa more steadily over the course of the day. And, like many people with Parkinson's, I will be exercising regularly, perhaps aided by the newest wearable sensor technology. Such therapies will surely buy me time.

It's sometimes said that a diagnosis of Parkinson's is not so much a death sentence as a life sentence. The issue for the people standing in downtown Grand Rapids that day, and for the millions of people with Parkinson's around the world, is what we make of that life sentence. The remarkable Parkies I have met on my journey—from Pamela Quinn to Tom Graboys to Brian Grant—have convinced me of the importance of maintaining a positive attitude, even in the face of scientific setbacks. For as Tom Isaacs once told me, "If you don't have hope in Parkinson's disease, you don't have anything."

NOTES

PROLOGUE

6 *except in Japan*: This anomalous finding so far lacks a convincing explanation, but the epidemiological evidence appears solid. A study by H. Kimura in Yamagata Prefecture, for example, found significantly more women affected than men (91.0 per 100,000 women versus 61.3 per 100,000 men). See H. Kimura et al., "Female Preponderance of Parkinson's Disease in Japan," *Neuroepidemiology* 21, no. 6 (Nov.–Dec. 2002): 292–96.

6 *prevalence increases with age*: While Parkinson's mostly hits the middle-aged and the elderly, about 10 percent of cases affect those under forty.

1. "DISCOVERY"

12 *We are in fact a rather impressive group*: After he was diagnosed in 2008, Flynn resolved to walk, run, bike, or swim ten million meters (over six thousand miles) by 2014. He achieved his goal by January 2014.

14 *In addition to producing medical discourses*: The ammonite *Parkinsonia parkinsoni* is one of several fossils named after him.

14 *Parkinson seems to have been*: His friend the paleontologist Dr. Gideon Mantell describes him as having "an energetic intellect and pleasing expression of countenance, and of mild and courteous manners; readily imparting information, either on his favourite science, or on professional subjects." A. D. Morris, "James Parkinson, born April 11, 1755," *Lancet* 268, no. 6867 (April 1955): 761–63.

15 *Ignorant of Parkinson's essay*: The collected *Briefe an eine Freundin* [Letters to a lady friend], written between 1814 and 1835, was published in 1909. Albert Leitzmann, ed., *Wilhelm von Humboldts Briefe an eine Freundin* (Leipzig: Insel-verlag, 1909). The letters are discussed in R. Horowski et al., "An Essay on Wilhelm

von Humboldt and the Shaking Palsy: First Comprehensive Description of Parkinson's Disease by a Patient," *Neurology* 45, no. 3 (1995): 565–68.

15 *He noted his stooped posture*: Humboldt gave a pretty accurate interpretation of why the task of writing was so difficult for people with Parkinson's. "Writing, if it has to be firm and clear, requires a lot of sometimes minute and hardly noticeable finger movements that need to be made in rapid sequence but with clear distinction with each other."

16 *According to the neuroscientist and historian*: Charcot examined his patients in front of hundreds of observers. Sometimes he asked the patients questions. Sometimes he meticulously inspected them for minutes on end, in absolute silence—students called this the "mysterious silence of Charcot." See Christopher G. Goetz, "The History of Parkinson's Disease: Early Clinical Descriptions and Neurological Therapies," *Cold Spring Harbor Perspectives in Medicine* 1, no. 1 (2011). He communicated any salient findings to his audience by acting out the symptoms, drawing sketches—at which he was a master—and using his patients as props. "If you tap on this patient, he will propulse forward and his gait will be quite unusual," lectured Charcot in 1887. "His head bends forward, he takes a few steps and they become quicker and quicker to the point that he can even bump into the wall and hurt himself." Quoted by Goetz, www.movementdisorders .org/monthly_edition/2009/08/charcot.php.

16 *James Parkinson's 1817 essay*: From Dr. Windsor, librarian at the University of Manchester.

16 *By carefully observing his own patients*: In 1872, Charcot wrote about the face: "The muscles of the face are motionless, there is a remarkable fixity of look, and the features present a permanent expression of mournfulness, sometimes of stolidness, or stupidity." Quoted in Christopher G. Goetz, Michel Bonduelle, and Toby Gelfand, *Charcot: Constructing Neurology* (New York: Oxford University Press, 1995), 119.

17 *And in addition to becoming*: Charcot's therapeutic ideas were more promising than James Parkinson's suggested treatments, which included bloodletting from the neck, blistering, and laxatives.

17 *He prescribed hyoscyamine*: Scientists later discovered such drugs block the neurotransmitter acetylcholine. These so-called anticholinergics were the only treatment available until the advent of levodopa therapy. Synthetic anticholinergics emerged in the 1950s. Charcot's British contemporary W. R. Gowers followed similar treatment strategies. For tremor, he used hyoscyamine and also noted arsenic, morphia, conium (hemlock), and "Indian hemp" (cannabis) as anti-tremor agents.

17 *His therapeutic vibration concept*: See A. S. Kapur et al., "Vibration Therapy for Parkinson's Disease: Charcot's Studies Revisited," *Journal of Parkinson's Disease* 2, no. 1 (2012): 23–27.

19 *But thereafter, all he had to guide him*: The gray matter, it would turn out, was made up largely of the neuronal cell bodies, and the white matter consisted of the fibers (axons) that connected gray matter (neurons) together. In the late nineteenth century, the Italian physician Camillo Golgi developed a breakthrough staining technique by soaking brain tissue in silver nitrate solution. The Spanish anatomist Santiago Ramón y Cajal used Golgi's stain to visualize the giant Purkinje nerve cells, revealing what appeared to be spaces separating one neuron's axon from an adjacent neuron's cell body. This suggested that the brain was made up of anatomically distinct nerve cells, or neurons. Ramón y Cajal and Golgi shared the 1906 Nobel Prize in Physiology or Medicine. It wasn't until 1950, however, that the tiny gap between two neurons—the synapse—could be visualized with the newly available electron microscope.

21 *The postmortem turned up*: The lump was an enucleated tuberculoma, evidence of tuberculosis.

21 *His dissertation examined fifty-four autopsied brains*: Six patients had conventional Parkinson's disease, and three were victims of the encephalitis lethargica epidemic that broke out between 1916 and 1926. Patients with this so-called von Economo's encephalitis (presumed to be linked to the 1918 influenza epidemic) had Parkinson's-like symptoms (parkinsonism), but there were major clinical and pathological differences. Unlike typical Parkinson's disease, von Economo's encephalitis came on rapidly and didn't progress. When this form of parkinsonism died out, regular Parkinson's disease continued.

23 *Tretiakoff also noticed*: The substantia nigra is damaged in both regular Parkinson's disease and encephalitis lethargica, but Lewy bodies were found only in the brains of people with regular sporadic Parkinson's disease.

23 *Tretiakoff named them "corps de Lewy"*: Lewy reported his discovery in 1912. He also reported strange inclusions in nerve cell processes (later called Lewy neurites).

23 *working in Dr. Alois Alzheimer's Munich laboratory*: In 1901, Dr. Alzheimer observed a fifty-one-year-old patient named Auguste Deter, who had short-term memory loss. After she died in 1906, Alzheimer had her brain examined and found the characteristic amyloid plaques and neurofibrillary tangles of the disease that now bears his name.

23 *Pathologists and neurologists would in time*: For decades, many scientists resisted the idea that Lewy bodies were a pathological hallmark of Parkinson's. It was only in the late 1940s, after the British neuropathologist Dr. J. G. Greenfield completed a large study of parkinsonian brains at London's National Hospital for Nervous Diseases confirming the presence of both nigral damage and Lewy bodies, that scientific attitudes changed.

23–24 *But only after pathologists*: *Stedman's Medical Dictionary* gives the following definition of disease: "a morbid entity characterized usually by at least two of

these criteria: a recognized etiological agent (or agents); an identifiable group of
signs and symptoms; consistent anatomical alterations."

2. RESTORATION

26 *On a hunch, he tried*: Like many molecules, dopa has two mirror forms: dextro-
rotatory, or D-dopa, and levorotatory, or L-dopa. It turned out that the L-dopa
form had far fewer gastrointestinal side effects.

27 *At the time, most neuroscientists*: Arvid Carlsson and his colleagues were the first
to recognize that dopamine was a neurotransmitter, an achievement for which
Carlsson was awarded the 2000 Nobel Prize in Physiology or Medicine.

28 *Just as L-dopa*: The conversion takes place in the terminals of the surviving dopamine neurons and depends on the enzyme dopa decarboxylase.

28 *"It was a spectacular moment"*: Oleh Hornykiewicz, interview with Barbara W.
Sommer, Feb. 9, 2007, Movement Disorder Society Oral History Project, www
.movementdisorders.org/MDS-Files1/PDFs/hornykiewicz.pdf.

30 *In 1966, in the first*: See C. Fehling, "Treatment of Parkinson's Syndrome with
L-dopa: A Double Blind Study," *Acta Neurologica Scandinavica* 42, no. 3 (1966):
367–72.

30 *With this new dosing regimen*: Ibid.

31 *It's easy to see why neurologists*: John Updike's *Rabbit Redux*, published in 1971,
mentions L-dopa. Rabbit's mother has Parkinson's, and Rabbit's father tells him,
"They've gone ahead and put her on this new miracle drug . . . L-dopa, it's still in
the experimental stage I guess, but there's no doubt in a lot of cases it works won-
ders . . . [H]er talk comes easier, and her hands don't shake that way they have so
much."

32 *But clinicians soon discovered*: By 1969, Cotzias was reporting not only dyskinesia
but also irritability, anger, hostility, paranoia, and insomnia. That same year, Yahr
reported that out of sixty patients, thirty-seven suffered from dyskinesia. Duvoisin
reviewed the data in 1974 and reported that six months out, 53 percent of patients
had developed dyskinesia, and a year out, 81 percent had. These clinicians also
noted motor fluctuations.

33 *"The patient on L-DOPA"*: Oliver Sacks, *Awakenings* (New York: Vintage, 1999),
247.

33 *"happy state—his world"*: Ibid.

33 *On the one hand*: The ten-year survival rate for Parkinson's patients increased
from 46 percent before L-dopa to 78 percent in 1974.

33 *You exchange moving well*: Evidence suggests that 40 percent of patients with
Parkinson's disease experience levodopa-induced dyskinesias four to six years
after starting the drug and up to 90 percent of patients get them after nine to fif-

teen years of treatment. See J. E. Ahlskog and M. D. Muenter, "Frequency of Levodopa-Related Dyskinesias and Motor Fluctuations as Estimated from the Cumulative Literature," *Movement Disorders* 16, no. 3 (May 2001): 448–58.

34 *About twenty people*: Epidemiological research validates this basic picture. In the general population, there are three men for every two women with Parkinson's disease. About 20 percent of cases present without tremor, as Charcot had observed.

4. MIND OVER MATTER

54 *"For anyone, learning that you"*: Pamela Quinn, "Moving Through Parkinson's," *Dance Magazine*, Dec. 2007.

55 *But I had read about some striking exceptions*: The video can be accessed at www .nejm.org/action/showMediaPlayer?doi=10.1056%2FNEJMicm0810287&aid =NEJMicm0810287_attach_1&area=&.

56 *Bloem, who has devoted*: See profile of Bloem in Neuroscientists' Corner, *Journal for Parkinson's Disease*, www.journalofparkinsonsdisease.com/JPD/Neuroscientists _Corner.html.

62 *So while it's tragic*: The Mayo clinician Dr. Jay Van Gerpen has developed the "Mobilaser," which attaches to a walker and projects a laser line onto the floor. By looking at the line, advanced Parkinson's patients can avoid festinating and gait freezing.

5. PATIENT POWER

63–64 *"In his opinion"*: Michael J. Fox, *Lucky Man: A Memoir* (New York: Hyperion, 2003), 21.

64 *"'You look like hell'"*: Ibid., 26.

64 *"Therapeutic value, treatment"*: Ibid., 29.

64 *"I can vividly remember"*: Ibid., 223.

66 *"my symptoms were still limited"*: Ibid., 202.

66 *"I've known for years"*: Ibid., 210–11.

66 *"It's made me stronger"*: K. S. Schneider and T. Gold, "After the Tears," *People*, Dec. 7, 1998, www.people.com/people/archive/article/0,,20127002,00.html.

66 *"There are so many things"*: D. Sloan, producer, *20/20*, Dec. 4, 1998, American Broadcasting Corporation.

66 *"Without intending to"*: Fox, *Lucky Man*, 228.

69 *"By the end of 1998"*: Ibid., 235.

70 *"We have a hearing today"*: *Parkinson's Disease Research and Treatment: Hear-*

ing Before the Subcommittee of the Committee on Appropriations, 106th Cong. 1 (1999), SH 9-28-99.

70 *"I concur that we are close"*: Ibid.

71 *"Will more money"*: Ibid.

71 *"If one of the congressional cameras"*: Michael J. Fox, *Always Looking Up: The Adventures of an Incurable Optimist* (New York: Hyperion, 2009), 228–29.

71 *"I think most of us"*: *Parkinson's Disease Research and Treatment: Hearing Before the Subcommittee of the Committee on Appropriations.*

71 *"If we can turn on even half"*: Ibid.

72 *"I did not graduate"*: Ibid.

72 *"I realized that I was"*: Fox, *Always Looking Up*, 228–29.

73 *"The last thing I want"*: Ibid., 54.

73 *"I need you to help me"*: Ibid., 55.

6. SURGICAL SERENDIPITY

79 *But Dr. Alim-Louis Benabid*: "Vim" stands for "ventralis intermedius nucleus."

79 *"The easiest explanation"*: See A. L. Benabid et al., "A Putative Generalized Model of the Effects and Mechanism of Action of High Frequency Electrical Stimulation of the Central Nervous System," *Acta Neurologica Belgica* 105 (2005): 149–57.

83 *"I'm no longer a slave"*: See J. Havemann, "Taming My Tremor," *Los Angeles Times*, Nov. 14, 2004, http://articles.latimes.com/2004/nov/14/magazine/tm -parkinson46.

88 *This, Alterman tells me*: The adjacent structure is the substantia nigra pars reticulata.

89 *Results of deep brain stimulation*: Meta-analyses of studies about deep brain stimulation to the STN find dyskinesias are reduced by about 70 percent, off episodes are reduced by about 70 percent, and health-related quality of life is improved by about 35 percent. But negative outcomes—diminished vocal function, increased drooling, cognitive problems—are also reported.

7. THE EXERCISE Rx

98 *Another research team*: See R. P. Duncan and G. M. Earhart, "Randomized Controlled Trial of Community-Based Dancing to Modify Disease Progression in Parkinson Disease," *Neurorehabilitation and Neural Repair* 26, no. 2 (2012): 132–43.

98 *Studies of boxing*: See S. A. Combs et al., "Boxing Training for Patients with Parkinson Disease: A Case Series," *Physical Therapy* 91, no. 1 (2011): 132–42.

98 *And epidemiological research*: See Q. Xu et al., "Physical Activities and Future Risk of Parkinson Disease," *Neurology* 75, no. 4 (2010): 341–48.

100 *RAGBRAI bicycle ride*: The acronym RAGBRAI stands for the Register's Annual Great Bicycle Ride Across Iowa.

8. NEW NEURONS FOR OLD

115 *Patrik Brundin resolved*: The family returned to Sweden, but in 1978 Patrik obtained a scholarship allowing him to return to the United Kingdom to attend the boarding school Atlantic College in Wales. He spent two years there studying for his international baccalaureate, for which he was required to write an "extended essay." The seventeen-year-old Patrik told his biology teacher that he wanted to do his project on Parkinson's research. The teacher referred Patrik to the school doctor, who passed him on to a neuropharmacologist, Professor John Davis, at nearby Cardiff Heath Hospital. Patrik told Professor Davis he wanted to develop an animal model of Parkinson's disease. His teenage logic was direct if ambitious. "I'd read that manganese miners in South America could get Parkinson-like symptoms from manganese poisoning . . . so I started in August 1979, feeding manganese chloride—that I'd dissolved in the drinking water—to rats. I came back in December to analyze them and found that they had developed movement problems." Patrik's article about his animal model, which he wrote by hand with an ink pen, was eventually published in a Pergamon press journal called *Project*, which highlighted some of the essays from Atlantic College every year. After passing his international baccalaureate, Brundin returned to Sweden to study medicine at Lund University in 1980.

116 *The entire fetus at eight weeks*: Strictly speaking, scientists use the term "embryo" for the first trimester and "fetus" for the second and third trimesters.

120 *One of the first*: See J. B. Moseley et al., "A Controlled Trial of Arthroscopic Surgery for Osteoarthritis of the Knee," *New England Journal of Medicine* 347, no. 2 (2002): 81–88; A. Kirkley et al., "A Randomized Trial of Arthroscopic Surgery for Osteoarthritis of the Knee," *New England Journal of Medicine* 359, no. 11 (2008): 1097–1107.

121 *The scrupulously polite Brundin*: The institute, named after the Grand Rapids entrepreneur Jay Van Andel, is housed in a striking building set into a hill. It sits on the edge of a medical area, which, thanks to a billion dollars of philanthropy, boasts a dozen world-class research and clinical facilities, including the Helen DeVos Children's Hospital, the Meijer Heart Center, the Lemmen-Holton Cancer Pavilion, the GVSU Cook-DeVos Center for Health Sciences, the Calkins Science

Center, and Spectrum Health Systems. There is so much concentrated medical excellence on display that Michigan State University chose to move part of its medical school there.

124 *L-dopa's first controlled double-blind trial*: As previously cited, Fehling examined intravenous administration of the drug in twenty-seven patients. Fehling, "Treatment of Parkinson's Syndrome with L-dopa."

9. NEUROPROTECTION

127 *Scientists have estimated*: See E. K. Pissadaki and J. P. Bolam, "The Energy Cost of Action Potential Propagation in Dopamine Neurons: Clues to Susceptibility in Parkinson's Disease," *Frontiers in Computational Neuroscience* 7, no. 13 (2013), doi:10.3389/fncom.2013.00013.

130 *Over the last two decades*: These include the DATATOP, TEMPO, and ADAGIO studies.

132 *In 1991, Frank Collins*: See L. F. Lin et al., "GDNF: A Glial Cell Line-Derived Neurotrophic Factor for Midbrain Dopaminergic Neurons," *Science* 260, no. 5111 (1993): 1130–32.

10. REBRANDING PARKINSON'S DISEASE

142 *"I was on top of the world"*: T. B. Graboys and P. Zheutlin, *Life in the Balance: A Physician's Memoir of Life, Love, and Loss with Parkinson's Disease and Dementia* (New York: Sterling, 2008), xviii.

143 *"A voice called to me"*: Ibid., 27.

144 *so-called stage 1*: In 1967, the neurologists Melvin Yahr and Margaret Hoehn published their scale of Parkinson's disease severity to map the natural progression of the condition. Patients advanced, they argued, from stage 1 (where the disease is confined to one side of the body), to stage 2 (bilateral symptoms), to stage 3 (involving postural instability), to stage 4 (impaired but still able to walk), to stage 5 (wheelchair-bound).

145 *There was also historical evidence*: For example, the dorsal motor nucleus of the vagus, the nucleus basalis of Meynert, the hypothalamus.

147 *Ross was able*: See R. D. Abbott et al., "Excessive Daytime Sleepiness and Subsequent Development of Parkinson's Disease," *Neurology* 65, no. 9 (2005): 1442–46.

147 *They looked first*: See R. D. Abbott et al., "Frequency of Bowel Movements and the Future Risk of Parkinson's Disease," *Neurology* 57, no. 3 (2001): 456–62.

147 *The researchers also looked at*: See G. W. Ross et al., "Association of Olfactory Dysfunction with Risk of Future Parkinson's Disease," *Annals of Neurology* 63, no. 2 (2008): 167–73.

147 *Other studies found*: See Abbott et al., "Excessive Daytime Sleepiness and Subsequent Development of Parkinson's Disease."

149 *"is where hope lives"*: See T. B. Graboys, "Finding Hope in the Midst of Despair: My Decade with Parkinson's Disease and Lewy Body Dementia," *Movement Disorders* 27, no. 11 (2012): 1358–59, doi:10.1002/mds.25117.

149 *But Tom, I realized*: Tom's aggressive type of Parkinson's disease—sometimes called Lewy body disease—came with the broadest package of symptoms, and postmortem examinations of such patients showed extensive spread of Lewy bodies not just in the substantia nigra but throughout the three nervous systems (see chapter 13). But the new thinking (elaborated in this chapter) argued Tom's condition wasn't really fundamentally different from mine; it just progressed faster.

11. THE DESCENDANTS

162 *who'd found a version*: See L. Maroteaux, J. T. Campanelli, and R. H. Scheller, "Synuclein: A Neuron-Specific Protein Localized to the Nucleus and Presynaptic Nerve Terminal," *Journal of Neuroscience* 8, no. 8 (1988): 2804–15.

162 *who reported isolating*: See K. Ueda et al., "Molecular Cloning of cDNA Encoding an Unrecognized Component of Amyloid in Alzheimer Disease," *Proceedings of the National Academy of Sciences of the United States of America* 90, no. 23 (1993): 11282–86.

162 *On May 27, 1997*: See M. H. Polymeropoulos et al., "Mutation in the Alpha-Synuclein Gene Identified in Families with Parkinson's Disease," *Science* 276, no. 5321 (1997): 2045–47.

164 *Spillantini had found the answer*: See M. G. Spillantini et al., "Alpha-Synuclein in Lewy Bodies," *Nature* 388, no. 6645 (1997): 839–40. Lewy bodies contain more than a hundred proteins. Spillantini says, "It's like a rubbish bin." But alpha-synuclein is not only the most abundant protein; it forms the filaments that make up the skeleton of the Lewy body.

164 *the legendary neuroanatomist*: The British neuroanatomist Chris Hawkes says of Braak, "I would rate him as the finest neuropathologist alive at the moment."

165 *Braak argued this was compelling*: See H. Braak et al., "Staging of Brain Pathology Related to Sporadic Parkinson's Disease," *Neurobiology of Aging* 24, no. 2 (2003): 197–210.

165 *Loss of smell*: Braak proposed the disease could start in the gut and then travel backward (retrogradely) along the vagus nerve to the brain. It could also start in the olfactory region and travel forward (anterogradely) to the amygdala. In stage 2, the Lewy neurites reach the brain stem (affecting the medulla oblongata and the pontine tegmentum, which manage sleep, swallowing, and autonomic functions like blood pressure). In stage 3, Lewy neurites reach the amygdala (smell, emotions like fear) and the substantia nigra. Says Hawkes, "These two paths [starting in

gut/starting in olfactory bulb] meet up in the temporal lobe of the brain. By that time, the patient is quite ill."

168 *"In Down's syndrome"*: In an earlier discovery of copy number variation, geneticists had found that a gene on chromosome 17 is duplicated in Charcot-Marie-Tooth disease (CMT) type 1A; CMT is named for Jean-Martin Charcot, his pupil Pierre Marie, and a British clinician, Henry Tooth.

169 *Hardy describes the news*: See A. B. Singleton et al., "Alpha-Synuclein Locus Triplication Causes Parkinson's Disease," *Science* 302, no. 5646 (2003): 841.

12. WHEN GOOD PROTEINS GO BAD

174 *A protein 100 amino acids long*: The human body uses 20 amino acids: arginine, histidine, methionine, threonine, valine, isoleucine, lysine, phenylalanine, tryptophan, leucine, alanine, asparagine, aspartic acid, cysteine, glutamine, glutamic acid, glycine, proline, serine, and tyrosine.

175 *He wanted to solve*: Chemists' favorite metaphor for how protein folding may work is a group of skiers descending a snowy slope. Skiers may choose different paths down the mountain, but thanks to gravity and the tapered shape of the piste, they all end up at the ski lift at the bottom. In the same way, proteins may start off folding in many different patterns, but however they start off, they are all forced down a predesigned "energy slope," and (without having to try every shape) they quickly—in a matter of a few millionths of a second—find their native conformation: the state of lowest energy.

178 *Prusiner quickly discovered*: Writing in 1772, the Reverend Thomas Comber vividly describes the clinical course of what would be called scrapie. A sick sheep, he observes, displays a "kind of high headedness . . . [and] . . . may appear wilder than usual. He bounces up suddenly . . . and runs to a distance as though he were pursued by dogs. In the second stage the principal symptom of the sheep is his rubbing himself up against trees, posts etc. with such fury as to pull off his wool and tear away his flesh . . . [In the] third and last stage . . . the poor animal appears stupid, separate from the flock, walks irregularly . . . and eats little."

179 *In the 1980s and 1990s*: Fatal familial insomnia is an extremely rare inherited disease that has been traced back to a man named Giacomo who was born near Venice in 1791. Victims suffer worsening insomnia leading to hallucinations, panic attacks, weight loss, dementia, and death. See D. T. Max, *The Family That Couldn't Sleep: Unravelling a Venetian Medical Mystery* (London: Portobello, 2008).

179 *Gerstmann-Sträussler-Scheinker disease*: In 1936, three Viennese neuropsychiatrists—Josef Gerstmann, Ernst Sträussler, and Ilya Scheinker—characterized this rare inherited condition in which victims suffer from speech difficulties, balance problems, dementia, and death.

179 *not just any proteins*: Prions are defined as "infectious proteinaceous particles."

179 *"It is difficult to convey"*: S. B. Prusiner, *Madness and Memory: The Discovery of Prions—a New Biological Principle of Disease* (New Haven, Conn.: Yale University Press, 2014), 93.

179 *In the mammalian body*: Prusiner benefited from the work of the British mathematician John Griffith, who wrote about this in J. S. Griffith, "Self-Replication and Scrapie," *Nature* 215, no. 5105 (1967): 1043–44.

181 *The secondary spread of fibrils*: Ice-nine is a fictional material appearing in Kurt Vonnegut's novel *Cat's Cradle*. Ice-nine is supposedly a polymorph of water more stable than common ice (Ice I_h); instead of melting at 0°C (32°F), it melts at 45.8°C (114.4°F). When ice-nine comes into contact with liquid water below 45.8°C (thus effectively becoming supercooled), it acts as a seed crystal and causes the solidification of the entire body of water, which quickly crystallizes as more ice-nine. As people are mostly water, ice-nine kills nearly instantly when ingested or brought into contact with soft tissues exposed to the bloodstream, such as the eyes.

182 *The results were published*: J.-Y. Li et al., "Lewy Bodies in Grafted Neurons in Subjects with Parkinson's Disease Suggest Host-to-Graft Disease Propagation," *Nature Medicine* 14 (2008): 501–3; J. H. Kordower et al., "Lewy Body–Like Pathology in Long-Term Embryonic Nigral Transplants in Parkinson's Disease," *Nature Medicine* 14 (2008): 504–7.

13. DAMAGE ASSESSMENT

186 *To date, the institute has performed*: From 1987 to 2005, the autopsies were done just on the brain, but from 2005 on the institute autopsies the body as well.

187 *According to Beach*: Also called neurofibrillary tangles.

189 *In the clinic, a skillful neurologist*: Even more confusing is the fact that LRRK2 mutations have variable penetrance. That is, the age when a particular genetic mutation will express itself (so that the individual will develop the trait, that is, parkinsonism) varies widely.

190 *The Lewy bodies and Lewy neurites*: Beach is a bit skeptical of the Braak staging system. "Braak has a staging system but not a grading system. In cancer, staging is the spread of the disease from site to site, and grading is looking at the actual microscopic appearance within any one site. Braak has a staging system—stages 1 through 6—but he never really incorporated a grading system, so it's possible to have one Lewy body in an area—which doesn't tell us anything about the disease severity—people don't get demented unless there's a lot of LB in the cortex, not just one. We should advance people from one site to the next only when we reach a threshold number of Lewy bodies."

191 *The researchers searched for signs*: Tyrosine hydroxylase (the enzyme that converts tyrosine into L-dopa) and dopamine transporter (a protein that mediates dopamine reuptake). See J. H. Kordower et al., "Disease Duration and the Integrity of

the Nigrostriatal System in Parkinson's Disease," *Brain* 136, pt. 8 (2013): 2419–31.

193 *If this finding*: Kordower's results make sense in the light of Paul Bolam's recent provocative work on substantia nigra dopamine neurons' energy conundrum. Bolam argues that these neurons are highly energetically vulnerable because of their extraordinary propensity for axonal branching. He estimates that a single unmyelinated dopamine neuron can have one to two million synapses and grow to an aggregate length of 4.5 meters. (This is orders of magnitude longer than other neuron types.) Substantia nigra pars compacta dopamine neurons, therefore, are energetically living on the edge. Anything that causes energy demand to exceed supply (genes, environment, age, and so on) sets them on a downward spiral where the axons degenerate and the terminals are lost. Interestingly, Bolam argues that using growth factors to promote growth and sprouting of surviving axons "may be counterproductive, in that it would increase the energy demands on individual neurons and hence their susceptibility." Beach concedes, "Neurotrophic factors will only work if you start super early. If you wait until the formal diagnosis has been made, there's often a lot of destruction, and you can't really hope to reverse that."

193 *"like drilling for oil"*: Uwe Reinhardt, interviewed for "The Other Drug War," *Frontline*, WGBH/PBS, 2003.

ACKNOWLEDGMENTS

I would like to express my gratitude to the many researchers, clinicians, professional colleagues, and fellow people with Parkinson's who helped me research this book.

I owe a particular note of thanks to Patrik Brundin, Andrew Lees, Bill Langston, and Dave Iverson, who read the manuscript in its entirety and offered valuable advice.

I would also like to acknowledge the School of Journalism and Communication at the University of Oregon for generously granting me sabbatical leave to work on this project.

During the writing of this book, I have had the privilege of getting to know many people with Parkinson's. I was saddened to learn of the passing of one of them, Dr. Tom Graboys, earlier this year. He was an inspiration to so many people and his memory will live on.

I owe a special debt of gratitude to my agent, Jill Kneerim, and my editor, Amanda Moon, whose encouragement and expertise kept me on task and on track.

And finally, my thanks go out to my family—Yanira, Catalina, Josh, and Mateo—for their unflagging love and support.

INDEX

Page numbers in *italics* refer to illustrations.